The Ellipse

The Fall and Rise of the Human Soul Secrets of the Cosmos

Journey into the Soul, Apocalypse, and Destiny of Humankind

A revealing of the spiritual science that is the source of all religion, and the cause of the human condition

by Zakariyya Ishaq

CCB Publishing
British Columbia, Canada

The Ellipse: The Fall and Rise of the Human Soul, Secrets of the Cosmos

Copyright ©2008 by Zakariyya Ishaq
Perfect Circle Communications
ISBN-13   978-0-9809995-2-5
First Edition

Library and Archives Canada Cataloguing in Publication

Ishaq, Zakariyya, 1953-
The Ellipse: The fall and rise of the human soul, secrets of the cosmos / written by Zakariyya Ishaq.
ISBN 978-0-9809995-2-5
Also available in electronic format.
1. Soul. 2. Cosmology. 3. Metaphysics. I. Title.
BD421.I85 2008   202   C2008-902831-7

United States Copyright Office Registration # TXu1-297-712
Cover image courtesy of Jet Propulsion Laboratory: www.jpl.nasa.gov

Extreme care has been taken to ensure that all information presented in this book is accurate and up to date at the time of publishing. Neither the author nor the publisher can be held responsible for any errors or omissions. Additionally, neither is any liability assumed for damages resulting from the use of the information contained herein.

All rights reserved. No part of this publication may be reproduced, stored in a retrieval system or transmitted in any form or by any means, electronic, mechanical, photocopying, recording or otherwise without the express written permission of the publisher. Printed in the United States of America and the United Kingdom.

Publisher: CCB Publishing
             British Columbia, Canada
             www.ccbpublishing.com

Dedication:

To all Humanity

THE FALLEN RACE WILL RISE AGAIN

Complete Human

Insanul-Kamil

Contents

Preface 1
Introduction 9
Glossary 13
Chapters Content
Chapter 1: The Fall of Adam 17
Chapter 2: The Perfect Circle and the Ellipse 29
Chapter 3: The Story of the Circles 47
Chapter 4: The Fall Continued 50
Chapter 5: The Cosmic Intention of the Evolutionary
            Intelligence (God) 67
Chapter 6: The Spiritual Hierarchy 73
Chapter 7: The Two Elliptical Lines of Descent 85
Chapter 8: History of Religion After the Fall 93
Chapter 9: The Ellipse and the Myth of Jesus 111
Chapter 10: The Six (7) Natures 119
Chapter 11: The Soul 130
Chapter 12: The Divine Epiphany 138
Chapter 13: Paradise and Hell 149
Chapter 14: Return to Perfection 159
Chapter 15: The End of Time: Perfection and Completion 179
ANNOTATIONS 183
Secret Knowledge
Number Places on Perfect Circle
The Secrets of the Cosmos
Unity of Faith

The Ellipse: The Fall and Rise of the Human Soul

Preface

This book, among other things, is a revealing of a primordial mystical science that has been lost from humanity for ages. Among the other things, it is the story of the cosmic law of the universe that rules the inner soul of man - the microcosm, and the outer soul of the cosmos - the macrocosm. I will tell this story using the metaphor of a *perfect circle* for the condition of the human soul when it was in a state of perfection, and master of the primordial mystical science. Antithetically, I will use the term *ellipse*, which means imperfect circle, as the metaphor for the soul as it evolved over time after its fall from grace - and losing the knowledge of the mystical science - as the allegory about Adam and Eve, recorded in the Torah and Quran, informs us.

The author will utilize this metaphorical dichotomy to elucidate this spiritual cosmology in order to facilitate understanding. Metaphors are used because they cover the full spectrum of levels of understanding that humans exhibit. They can be read on a basic level, such as reading the story of the Garden of Eden as a simple moral tale of obedience. But they can also be interpreted on a high level, as here, in a way that tells us about the structures of the soul, that the allegory is informing us about.

I could have used the term paradise, instead of circle, but this brings too much emotionality into the story, I feel; therefore the more objective term "circle" is being used in order to create a unity of understanding, devoid of the emotion attached to a word generally used in a religious context.

The ellipse is the structure that the soul of humans has evolved to - what I call imbalanced balance! Opposed to this is its opposite, the perfect circle, the universal symbol of perfection - the inner and outer world of felicity we abided in before the fall, as told in the allegorical scriptural tale of Adam and Eve in the Garden of

The Ellipse: The Fall and Rise of the Human Soul

Eden. Humans lived in harmony, peace, and happiness in this period. Then something terrible happened, something that literally altered the nature of humans, and consequently produced the imbalanced yet grotesquely "functional" world of the ellipse, the imperfect, perverted world we have lived in since this ancient trauma.

The basic premise of the ellipse theory of history is that the fundamental reality of existence revolves around the world we occupy. The world in itself and the beings who occupy this world, it turns out, are only microcosms of that larger world that can be called the macrocosm, or the world in totality, as opposed to the microcosm - each individual occupying that world. This wisdom is from the tradition handed down to us from the great teacher Hermes known by many different names to different cultures. In a similar vein to the tenets of Taoist philosophy, he taught us that man is but a microcosm of this greater world.

This is absolute cosmic law, not any arbitrary "creation" of humans and the world by a "God". This creation may be the mechanism in which the transcendent cosmic law is the main determiner of, nevertheless if so, then the law is as important, or more so, to the divine epiphany as the creative agency.

This fundamental concept is the most vital tenet in understanding genuine metaphysics and spiritual cosmology. Indeed, whatever worlds exist anywhere, this law is an existential reality that will always operate on the same principle as it operates in our world. Therefore it is the first thing we must come to grips with in understanding the inner and outer universe and how it works, and how we are to gain control of it for the benefit of all humankind.

The ellipse, the metaphor for our fallen world, designating the intrinsic corruption that we have evolved to since the fall, is the world that shapes all of us, and determines how we feel, how we live, and is the source of the experience of all in the world. The ellipse has a source, and that source is the perfect circle - its dichotomy - indeed, we learn through this that the source of hell is

Preface

paradise.

The concept of the macrocosm shaping the microcosm transcends the ellipse concept. In other words, if we did not fall in the abyss, the cosmic reality that microcosms are always shaped by macrocosms (and the other way around in some instances) is intact. If we were still perfect circles, as in primordial times, then the outer world-macrocosm would be perfect. Unfortunately for man, we have evolved into a corrupt imperfect circle inwardly, and the macrocosm, in this case, has followed suit.

Our world shapes us, creates us, and determines our experience. This is unavoidable - no one can escape this fact. Certainly, there is a great degree of variable difference in our experience, for example between the rich and the poor. This variation though does not go too far: the rich may face the same sickness and mental suffering in life as everyone else, and the inevitable pangs of death like any poor man. Certainly the parameters of this challenging existence is pretty much understood by us now, and therefore any extraordinary major change or difference has never been widely seen - if it exists - and will certainly be easily identified if it reveals itself. We know that some are rich and live pretty well relatively; some are secure materially, albeit not rich; but we also know that there are poor people whose suffering is acute. Despite the relative comfort of the few, in the entire world none will escape the inevitable periods of psychological and physical suffering that all humans face. So, we see the parameters of the degree of suffering is not that far apart.

The ellipse is not a mental illness in the clinical sense; of course it includes mental illness, nevertheless the ellipse just as much enslaves a person who is not mentally ill, as one who is. It is the basic flaw in our inner self that allows all the imbalances that cause inevitable disharmony, suffering, evil, ignorance, and wayward desires that have plagued us since the fall.

Moreover, the ellipse is also the cause of ordinary spiritual blindness, the opposite of the spiritual phenomenon known as awakening. This blindness is rooted in the conditioned self, another symptom of the ellipse.

This is because the world that makes up our reality has become an ellipse. (In this instance of the polarity - microcosm and macrocosm - there is a reversal of the usual phenomenon, in that in this case, the microcosm created the elliptical macrocosm.) This is the sole etiology of human suffering, going beyond the ideas of Buddha to the exact root of this problem, beyond his assertion that excess or wayward desire is the cause of our plight in the world. Desire is not the real root of our problems because wayward desire, like any psychological trait, has a spiritual root sourced in the higher structure of the soul. Beyond this concept, we understand that such psychological traits in humans exist based on a fundamental milieu such as the chemical environment in the brain. Hidden above or beyond this chemistry is a hidden world of celestial realities that rule this lower world of chemistry and the like, where "desire" takes root. What actually happens is that the condition of the higher realm determines what exactly will occur, by reflection in the lower psychic realm, all the way down to the physical chemistry.

This metaphysical phenomenon is based on the concept of the four worlds:

SOUL
SPIRIT
MIND
BODY

Humans since the fall have lacked cognition of the two higher worlds, therefore do not have any information about the ellipse that exists there. Most often, through the practice of metaphysics, does one regain consciousness of the higher realms and the possibility of resolving the ellipse.

It must be said that the information revealed in this book will not guarantee anyone anything unless they do something for themselves, through action. This is not a how-to book, nor a magic bullet. However, this exciting, dynamic, and new cosmology will offer a genuine roadmap to the theoretical liberation of the human being. Additionally, and more importantly, I will emphasis that no one is special in the eyes of God because of the circumstances of their condition or birth. This is vital to understand because it is intended to explain in this document that there is an elite group of beings operating in the world. And lest anyone take this to mean that they are somehow a part of this group - as many religions espouse - because of what they believe or think they know, they will be advised to relinquish this notion. Ironically, those who concentrate on this "superiority" complex are in fact demonstrating that they are not a part of any elite.

The revelation of this cosmic science is the final paradigm that explains the reasons for all of the circumstances that relate to suffering, evil, and the destiny of our race.

As the theory unfolds in these pages, some may become offended that their "God" seems excluded from this formula. Let it be known that whether or not there is some kind of creative agency that produces humans, they nevertheless are bound as we are by this cosmic science, and can do nothing without being within the dictates of its law. In addition, any divine agency must obey the dictates of reason, and the basis of real science: the law of cause and effect. Not to mention that ALL religion is basically the same in theory, just differing in nomenclature.

"God" does have a reality, but in the tradition of the Sufi and Buddhist cosmologies, even God is somewhat irrelevant in this cosmic drama on one level, in that he is bound by his own laws. Even he cannot do anything outside of these laws, though on another level, he is all and everything, he is the law, in that sense. Whatever the ultimate reality of God is in relationship to us, one thing is for sure, any assistance he renders man is only in the context of him being obligated to a higher law that binds himself to himself to do things by the law and within it, not the other way around, as traditional exoteric religion has suggested. He, God, then is the law by this logic. His specialty is what is termed "Evolutionary Intelligence," epochal knowledge of the minutiae intelligence man evolves to over time that necessitates new guidance for the new era.

The God principle will be dealt with by explaining to the reader that for millennia, humans have been misled about the concept of God. They have, through exoteric religion, been programmed to view God as a Santa Claus, Wizard of Oz figure having nothing to do with cosmic law or science. Therefore, it is a section "The Divine Epiphany" on how to view the divinity in the correct prospective. I will not dwell too much on this because the subject of the book is too important to go on in that vein that only elicits controversy; the goal here is to instruct and enlighten.

As I come to explain this essential reality that rules all of our experience of this world's life, and understanding is planted in the reader, we will consequently see the source of our extrication from the suffering that holds us in its grip. It is in our hands at one end of the spectrum of reality, and in the hands of time, at the other end.

This spiritual cosmology will elucidate clearly from the perspective of the highest world - inwardly, and outwardly - the science and mechanisms of the human soul. Or in non-mystical language, ordinary human psychology, feeling, and its source. This will go beyond the conceptual, to the literal physicality of the soul, and the theoretical delineation of its content.

Additionally, generally and specifically, it will describe humans' psychic past and future on this plane of existence, and why and how evil and suffering came into being, and how it will end in time. It will be demonstrated that the ellipse is the ultimate source of all evil, and without the ellipse, evil and suffering could not exist.

On one level this story will appear to be negative to the positive thinkers who want to "wishful think" humanity out of its doldrums; as if such antics will work, and scientific realities do not matter. For them, they can be rest assured that this cosmic drama will end with a positive outlook, based upon the cosmic scientific reality that I reveal and explain, not any wishful or positive thinking.

Based on the experience of gnosis and study of this topic for three decades, I began work in this field by studying intensely certain

Preface

religious, spiritual and occult themes that led to the synchronistic meeting with a timeless spiritual Adept who I believed at the time was St George {Khidr}. This was the beginning of my search for the knowledge that rules the cosmos. St George Khidr is the adept who rules an aspect of the world that deals with religion, politics, mysticism on an exoteric basis, among other things.

Primarily understood and explained through the myths of Western and some Eastern religious mythology, which includes mainly the Judeo-Christian-Islamic Western tradition in addition to Hermetic lore and science. The particular themes in the Quran and Torah that provide the mythical allegory of the fall of man are the basis of the ellipse theory. The allegory of Adam and Eve will be fleshed out of its mythological content and interpreted literally on a scientific, structural plane where we can clearly see and understand where we are in the cosmic mystery of the self, and why and how the ellipse has evolved to be the source of our suffering.

The ellipse theory answers many of the age-old questions that have challenged the understanding of believers in God, philosophers, and mystics, by intimately exploring these topics:

The essence of man

The origin of evil and suffering

The end of evil and suffering,

A definition of God

The reality of creation

The nature of the soul

The burden of the apocalypse

A new understanding of the Jesus mystery

Finally, and most importantly, a clear and precise delineation of the inevitable good destiny of all of humankind based on the science of what is revealed in the book. This science is universal, unalterable cosmic law that in the end will give all humanity: A Good Old Destiny!

Zakariyya Ishaq,
May 2008

Introduction

Imagine a world not much different from ours outwardly, but looking closely at it, it lacks very important things in it that we have. It does not have violence, ignorance, crime, wars, sickness, old age, death and other terrible things we spend all our time fruitlessly trying to eliminate. We are so used to these elements in our world that we cannot even conceive of any other possibility of a world devoid of these negative elements. Even the rich and wealthy in the end will face sickness, death, and suffering. Many of them also live with the same psychological suffering as any of us - despite their wealth, that in the end will not save them from what the great teacher Buddha saw in the decrepit old man wasting away who inspired him to seek the knowledge of what causes this inevitable suffering. The question then arises: What can we do to understand and possibly eliminate this condition?

According to tradition, Buddha eventually found the meaning of life, and the etiology of suffering. This condition he labeled *dukkha*, a Sanskrit word denoting suffering. The first noble truth in his system of spiritual cosmology was that humans are in an existential condition of suffering. This suffering he said has a cause, and that cause is ultimately desire, that eventually will wreak inevitable havoc with all of our lives. By desire, he meant excess, wayward, or perverted desire that denotes internal imbalance somewhere in the soul. His design to free us from this condition is enumerated in the rest of his philosophy. According to him, if one can follow his well-defined philosophy, this will make existence bearable as well as enabling you to acquire peace, happiness, and enlightenment.

He taught that there was a cause of everything, and that if one examined life closely one could find the cause behind any phenomenon.

Many of his students questioned him about deep metaphysical

topics, such as, what is the nature of God? Often he answered with silence, or he would offer the following analogy to the enquiries: If you find yourself with an arrow stuck in your heart, should you waste time and contemplate what kind of arrow is in you, or rather, wisely devise a method for taking out the arrow before it poisons you to death?

One can't help appreciating such wisdom, and cannot argue with it. However, I say, what if your child had an arrow inside of him in the back yard of your home, and you took it out and saved him? Would you then send your other child to play in the same back yard without first finding out who or what is shooting arrows at children? Far be it from me to dispute one of the wisest of men. However, is it not the nature of we modern thinkers to seek the knowledge even beyond the prescription of the Buddha, since indeed whatever he discovered, it has not remotely eliminated suffering on this planet, and his prescription carries a price most cannot pay to arrive at his success?

Most humans find it impossible to achieve that prescription of holiness, perfection, and intense meditative practices, along with the sincere practice of virtues. In addition to that, it has dawned on many that the Buddha as a philosopher, amongst many of his attributes, was most likely referring to beginners in his system - who should shun seeking answers to philosophical and cosmological questions. And not to established practitioners who he knew full well would seek answers to the question of the meaning of life, as he himself did.

Suffering, according to Western spiritual cosmology - the lore of the "People of the Book" that includes Judaism, Christianity, and Islam - began in the era of pre-history known as the primordial period, when humans once lived in a world where suffering was non-existent. The ancient scriptures provide an allegory for these events that describe our sojourn in a paradise on earth. That, according to other traditions outside of Western spiritual lore, was a period of earthly bliss in our pre-history that encompassed millennia; in fact, one could not even measure the period since time itself, as we know it, did not even exist.

Many traditions in religion, from the lore of the shaman of the Americas to the ancient Egyptian and Sumerian religious texts, record a time when man lived above the sufferings he is steeped in today.

The ancient lore of the mythical Atlanteans, and Lemurians, whom historians still today dispute, is another historical series of legends that have pervaded our lore of history that speaks to this mysterious period of pre history that intrigue us even today, as in the past.

Were these legends true concerning the ancient Atlanteans and the Lemurians? Many claim that they are - and have offered proof - although some of these individuals are called "alternative historians". Nevertheless, it seems certain that there is a pre-history that is unknown to us, and what we know of it is only from the holy books that many of us accept at face value. This mystical and religious lore cuts across many traditions over the spectrum of all of our culture's myths, and stories that indeed tell us something. Many people feel there must be a reality behind these ancient myths that seem to pervade our entire race, and the same or very similar myths exist in divergent peoples all over the globe. The story of the Garden of Eden or a paradise on earth is widespread, and seems to be an indigenous myth in many cultures' religious lore. It is the opinion of this author that the tradition in our heritage that has come down in history, concerning the great agility, and awesome power of the Atlanteans and Lemurians, is an indication that a part of their race is probably the race of "devils" or jinn, that was partially responsible for the fall of man. The myths known to us about them, though, are accounts of this archetypical period of the race of superiors recorded in *The Book of Enoch*. Although they cannot be blamed exclusively for man's fall, they did have a hand in it.

The minute details of who is who in history ultimately does not matter, since this is pre-history. The important thing in this entire chronicle is the science of the cosmic law and events that occurred on an inward level; this book minutely details that inner, hidden history.

The allegory of the Garden of Eden, and the fall of man, as recorded in the Torah and Quran, is about the inner psychic elements of humankind that got distorted through a horrible ancient experience that is based on a spiritual science of states and stations. The story conveys the events of how humans became trapped in a lineage, tree, or vine of one of these states.

Because of the fall, the mystical and religious sciences - according to metaphysical lore - were revealed to humans as a guidance to return them to the nature they lost in the Garden of Eden. The Quranic theme of "That all will return to us" is the cosmic law that determines this reality and its inevitability.

This book is not a linear history. It is a revealing of the spiritual cosmic laws that determine the quality of our existence. Certain psychic-spiritual elements went awry in the human soul because of the fall. This occurred due to inner cosmic laws, not any curse, or overt act of punishment by any sky-God. The elements that were corrupted are from the very essence of our inner soul. These elements are slowly returning to their original position in the soul structure of all human beings. Indeed, it is that simple in the end, as if one would go outside and skillfully sling a boomerang in the sky, and just as that boomerang in a certain time will return to the thrower, similarly these awry elements of our transcendent soul are on a trajectory of return to us.

So our story begins with that great event in pre-history, that all of us deep down in our souls know all about, yet do not seem to remember. But as recorded in these pages, we indeed remember what happened, on a very deep level, and know what will happen in the end.

The end of time.

Glossary of Significant Terms

Atlantis and Lemuria: Primordial civilizations of supernatural beings known as jinn. They were immensely powerful, whose activity grounded the solar power of the universal energy grid of the world for the benefit of the greater lot of humanity. From this great race came a group whom were the "devils" that deceived the lower Adam in the Garden of Eden.

Circle: Term used as a metaphor for the essence, tree, or state of consciousness that can be inter-dimensionally emanated in eternity. Within these states are incredible conditions of bliss. Additionally, the term is used as a part of the metaphorical dichotomy of the Ellipse-imperfect circle - hell, as the opposite of the Garden of Eden-perfect circle - paradise.

Circular Traveling: Term that describes traveling inwardly into the essence states of paradise (Garden of Eden, perfect circle). This can also be referred to as eating fruit of the tree (scriptural metaphor) or traveling in the vine of an essence or circle.

Dharma: Transcendent law of the universe.

Ellipse: Means imperfect circle. This is the metaphor used to designate the fallen condition of the soul of man.

Essence: Spiritual technical term for our proverbial circle and Garden of Eden. The essence is the face of the soul and one of the three transcendent aspects of it that essentially determine our state of consciousness.

Evolutionary Intelligence (EI): Term for God in history, who knows the needs of the evolving human, and consequently through revelation guides man through varying epochs.

Garden of Eden: The metaphor used by the scriptural authors for the perfect circle, or in spiritual technical terms the essence.

Holistic Self: The seventh nature, and the term that designates the ultimate completed perfection of God in form, the highest holon.

Holon: The existential reality of the entity, that is a whole in itself as well as an aspect of greater phenomena of similar entities, called a holarchy. There are sentient and non-sentient holons.

Holy Spirit: Aspect of the soul that nourishes the essence, and where all attributes of the active God reside.

Iblis: Term for the devil that connotes separation from the whole.

Insanul-Kamil: Completed soul or person that is the highest condition a human can reach; the very intent of God.

Jahanum: Arabic word from the Quran that means hell. This word is derived from the word for paradise, Jannah. Jahanum is essentially the spiritual technical term for the ellipse.

Jannah: Quranic term for paradise. Jannah is a Garden, body of celestial lights, Garden of Eden - paradise. It is also a generic term for the essence.

Perfect Circle: The metaphor for the opposite condition of the ellipse. The perfect circle is the condition of the soul of man when he was happy and enjoying paradise before the fall.

Rau: The term for God, that means unlimited free infinite spirit, energy, life, being, and non-being. The Rau is the source of all and at the same time completely free form all form.

Six natures: The six transcendent natures of reality, that exists in all sentient beings, on all levels.

Glossary of Significant Terms

Soul: The ultimate energy consciousness machine of man, which is the source of his being.

Tao: Unique entity of the soul that relates to the infinite Rau.

Vines: The metaphor for climbing the tree or circle lineage of paradise consciousness.

Chapter 1

The Fall of Adam

"He began the creation of man from dust. Then He made his progeny of an extract of water held in light esteem. Then He made him complete and breathed into him of His spirit and made for you ears and eyes and heart."
Quran 32: 7-9

"And We said: O Adam! Dwell thou and thy wife in the Garden, and eat ye freely (of the fruits) thereof where ye will; but come not nigh this tree lest ye become wrongdoers."
Quran 2: 35

"And the LORD God commanded the man, saying: Of every tree of the garden thou mayest freely eat; but of the tree of the knowledge of good and evil, thou shalt not eat of it; for in the day that thou eatest thereof thou shalt surely die."
Genesis 16:17

In primordial times before the fall of Adam, in the great civilizations know as Lemuria and Atlantis, this classical period reflects the era when the great civilizations of the jinn were flourishing as recorded in *The Book of Enoch*, the Torah, and Quran. In these two books, though, the identity of the "devils" (in Islamic cosmology a group of jinn) is disguised. Also, the tale in these scriptures is an allegory, with no intention to identify the parties in the story.

In Islamic cosmology, the jinn are the aspect of sentient being that the "devil" is derived from. Certainly, the devil was no angel, but of the race of powerful beings called jinn who are somewhere between the level of ordinary man and the extraordinary beings known as angels.

The Bible and Quran record an allegory that describes a "first" man, Adam, and a woman, Eve, in a Garden of felicity called the Garden of Eden. God tells them "You may eat of any tree in the garden you wish, but approach not this tree." The stories in each of the scriptures are slightly different: in the Bible, it says that the tree is the tree of good and evil. In the Quran, it is called the tree of immortality. The tree was not a physical tree nor was the garden a physical garden, for this is an allegory, a very high-level one that is referring to an aspect of the soul known as the *Essence*. The essence is the Garden of Eden; a spiritual structure that has to do with the invocation in human consciousness of spiritual states and stations of a lofty order, as well as states of indescribable bliss. This will be explained more in depth as we go along.

It is contended here that a small group of the people of Atlantis and Lemuria are the beings referred to in the scriptures that deceived Adam and Eve in the Garden of Eden. This identifies a group of this powerful psychic race as the "devils" in the Bible as well as the "jinn" in the Quran. The subject of this book is not a historical chronicle, however, and this argument is only presented for unity and clarity. This point is not vitally important in elucidating this spiritual cosmology, in which the vital aspect has nothing to do with linear history, so those whom wish to argue this point, hold your breath, and wait and hear out the rest of this story.

This early period is the key to understanding the fall of Adam and his creation. That is his creation as the fallen man from the perfect uncreated essence of the perfect man: the primordial Adam, or transcendent Adam, "Our Father in Heaven" as Jesus identified in the New Testament. There were two Adams: the transcendent "Father" and his son the lower Adam. This concept will be fully explained as we go along.

The Fall of Adam

"He began the creation of man from dust. Then He made his progeny of an extract of water held in light esteem. Then He made him complete and breathed into him of His spirit and made for you ears and eyes and heart."

Quran 32: 7-9

The historical legends about Atlantis and Lemuria, and the interpretations that go along with these two periods, are varied and many. From the occult oracles of the American mystic Edgar Cayce, to the mundane and historical expressions of Plato; or from the redoubtable legacy of Madame Blavatsky to the arcane ideas of the contemporary philosopher and New Age thinker Michael Tsarion, we have a wide variety of interpretations and explanations of the mythical lands of Atlantis and Lemuria. In fact, this period, well documented in the Quran, Bible, and *The Book of Enoch* on an allegorical level, is the period when the great jinn cultures were holding down the cosmological energy of the sun to the benefit of all humanity. Indeed, these high-powered energetic civilizations of immense psychic energy were processing not negative energy, but very high-powered energy from the solar sphere, strictly in accordance with the Evolutionary Intelligences (EI) (God's) program to initiate a vicegerent on earth - that is, Adam (lower). That vicegerent would in fact be a balancing archetypal apparatus in the forming of the essential being known as man. This is relative to the interior being that is made up of six primary inner natures, plus the seventh nature, or nature of the holistic self, which embodies the six natures:

1. Divine
2. Adamic (Higher/Transcendent)
3. Angelic
4. Adamic (lower); the new creation
5. Jinn
6. Anima

(7) Holistic self

The creation or maturing of the lower adamic nature was coming to fruition while the great jinn cultures were in their heyday. As the Quran records the epochal drama, it tells us that the EI (Evolutionary Intelligence) expresses to the existing developed natures of the angelic, jinn, and transcendent Adam that it intended to develop further the Adamic-lower nature in the existing mix.

Quran 2: 30 "And when thy Lord said to the angels, I am going to place a ruler in the earth, they said: Wilt Thou place in it such as make mischief in it and shed blood? And we celebrate Thy praise and extol Thy holiness. He said: Surely I know what you know not."

It expresses the intent and reminds the other natures that their duty is to continue in their responsibilities (that is, maintaining the balance of their respective light energy sources). Thus making way for the new Adam, who for a time would be the focus of the other nature's attention, considering that this is in a sense a new, or emerging, immature nature that needs special attention. Translated from this allegorical reference to spiritual, solar, and alchemical levels, the saga reflects the emergence of the new Adam, as yet unknown to the other natures, but known to God as the Quran reflects: **"Surely I know what you know not"** concerning the other natures' attitude about the new Adam.

History records that the jinn nature was not enamored to say the least of the new situation. Alchemically, it is in fact only a reflection in the holistic being, man, having a period of adjustment in the mixing of the developing natures, particularly the mixture of the emerging lower Adamic and the more mature ripe and fiery jinn strain. This is the classic conflict between the lower Adam - the earthly, claylike psychic substance - versus the fiery emotion-laden high-energy psychic substance of the jinn nature. The Quran addresses this problem in the volatile mixture of energies when it expresses the statement to the angelic and jinn strain to make obeisance to Adam:

Quran 2: 34: "And when we said to the angels, Be submissive to Adam, they submitted but Iblis (did not). He refused and was proud, and he was one of the disbelievers."

Later it also says:

"He said: What hindered thee that thou didst not fall prostrate when I bade thee? (Iblis) said: I am better than him. Thou createdst me of fire while him Thou didst create of mud."

Quran 7: 12

We know that in the great evolutionary design there is a problem with the jinn (FIRE) energy and lower Adamic (DUST) energy; we know that by the allegory of the scriptural traditional Biblical and Quranic tale of the "devil" (jinn) energy not submitting to the lower Adam or opposing him. Only in time as our enfolding history reflects will these seemingly opposing natures settle down and operate smoothly as the Evolutionary Intelligence (God) intended.

THE SIX NATURES AND THE FALL, CONTINUED

The six natures of being are the emerging evolutionary mixture of the micro-macrocosm of the God being, which includes the jinn strain (that has been incorrectly referred to by some clerics as demonic nature) as well as the other natures. The jinn strain is as important to the holistic being as any of the other natures. The primordial problem of the "devil" or jinn verses man is only in reality an alchemical problem that in time will go away as the settling of the volatile mixture of jinn and lower man settles down as we develop. In other words, in time the holistic human, that is all of us, will become balanced, as the six natures become used to each other. Additionally, as in the current aspect of the holistic being we are viewing that being - the lower Adam - versus our jinn nature. The other natures (divine, higher Adam, angelic, animal) albeit more attuned to the lower Adamic energy nevertheless require a period of adjustment.

It is essentially an inner conversation going on between the divine nature in union with the intelligent evolutionary attributes of God moving down to the transcendent human nature - higher Adam - to the angelic, to the jinn, lower Adam, and to the anima nature.

Therefore, we have in the allegorical tale the EI expressing its desire that the other natures make obeisance to Adam. This is not a literal command for the other natures to make any kind of gross acknowledgment of the lower Adamic strain's superiority, but is only an intention of the EI to make the lower Adam (feelings of well being) the focus of primary consciousness, or the axis of awareness of the essential emergence of the complete human being. The devils' "rebellion" therefore has to do with the jinn nature competing with the lower Adamic nature for the attention of primary consciousness or feeling. This is borne out by the experience of all human beings in their inner struggle between the turbulent simulative nature of our jinn selves as opposed to our lower Adamic selves, primarily concerned with well-being, peace and safety.

"He said: What hindered thee that thou didst not fall prostrate when I bade thee? (Iblis) said: I am better than him. Thou createdst me of fire while him Thou didst create of mud."

Quran 7: 12

The Fall of Adam

The allegory continues with Iblis not making obeisance to Adam. In the historical dimension, Iblis not making submission to Adam records the time when a faction of the great jinn civilizations of Lemuria and Atlantis had received understanding from the EI, that their high-energy-based civilizations were shielding "the sun's heat" from the new emerging Adam. The lower Adam's consciousness and heart had to be shielded from the high-powered energy of the universe; at least until our little Adam could be acclimated to his new reality. (Remember this is not a new being, but the emergence of the latent lower - Adamic - nature becoming manifest.) Thinking that one's activity, however useful, is basically designed to be at the service of some other aspect of the cosmic order was unacceptable to the very proud and occult-adepts of this particular group of the ancient Atlantean and Lemurian Demiurgic hierarchy. Therefore, as recorded in The Quran, Genesis, and more particularly in *The Book of Enoch*, they rebelled against the cosmic intention of the EI.

What must be understand today, as the legends of the power, glory and awesome agility of these ancient beings, and their civilization records, is that those who led the rebellion against the intention of the EI, the powerful adepts (Iblis in the Quran, Satan in the Torah, Azazil, in the book of Enoch) that were an integral part of the cultures of Atlantis and Lemuria were immensely powerful psychic beings. They had to be; for imagine the scientific, spiritual and occult knowledge one has to have to hold down the cosmic power of the solar heat! Consequently the rebellion against the EI, although foreseen and anticipated (and in-fact occultly encouraged by the EI, which I will cover more in depth later), was an immense event in cosmic history, a terrible traumatic drama, which human beings have groped from time immemorial to understand.

These great events whose knowledge has come down to us in these allegories in the Bible, Quran and the other Hindu, and Sumerian mythological scriptures, are in fact allegories that indeed are describing the creation of man archetypically and alchemically.

These stories should not be taken entirely literal. In fact, their relationship with literal history is an approximation at best. To understand this one must see that for instance, when the Quran talks about Iblis not obeying God's command, it must be understood that these events took centuries or millennium to manifest in the world, and in the soul of humans. In particular, the result of his rebellion took centuries to manifest. Consequently, a simple verse in the allegorical chronicles of psycho-spiritual history can be talking about hundreds of years. That should always be kept in mind, so the mistakes of the past, by primarily literal-minded clerics, can be avoided.

One of our greatest errors of thinking in religious theology, cosmology, and philosophy, is in not understanding that beyond the basic precepts that cover our actions towards each other, such as the Ten Commandments, scriptures are allegories that document - from the experience of advanced beings - how the inner and outer universe works. The stories and tales have primary value as allegories, and mythology. This has been forgotten by religionists, and misunderstood by elitist philosophers, who do not realize that the writers of scripture had to communicate on all levels, so the simplest believer all the way to the highest philosopher could understand things on their respective levels.

This is not to say that allegorical or mythological history has no basis in true history, not at all, the point is to avoid literalism and linear thinking, so as to foster understanding and avoid dogmatism and coercive thinking. As in this interpretation, I wish to emphasize, on one level, the importance of the story as it relates to humans psychologically and alchemically, since I am claiming here that the natures of the human are all important, and should be looked at holistically, while understanding the deep nature of the unity of being, and the universe, with integration in mind rather than separation, which is the essential enemy of human understanding and growth.

Therefore, Iblis on one level, that is the level of the psychic self, rebelling against God is one aspect of the nature of man that is the jinn nature, having a temporary conflict with the lower Adamic nature.

"And when We said to the angels: Make obeisance to Adam, they did obeisance, not so Iblis, he refused and was haughty and he was of the: rejecters."

Quran 2: 34

This haughtiness is only reflective of the powerful jinn nature as opposed to the lower Adamic nature, being made of clay, dust, or of a meek disposition.

THE JINN

This can be understood by all in reflecting on the universal conflicts in ordinary life that come into play in our experiences in life. Indeed, the lessons in the Sunday school primer, or the madrassa instruction about human temptation, reflect this simple, timeless conflict in human nature; this is not lofty metaphysics. On the other hand, the traditional clerical interpretation of this inner-psychic conflict has given us more fire than light on the subject. Certainly, the classical "devil" given to us by the clerics is not the complete story in trying to understand human psychology. For they often neglect to tell us about the other side of the "devil" besides the traditional devil of debauchery, that any kid understands. They neglect to tell us about the subtler devil of indoctrination, dogmatism, narrow thinking or the "creeping knave" of Blake, personified in fact by the cleric himself at times; in essence, the devil of the conditioned mind. So in other words, the ancient conflict played out in the macrocosm and microcosm over a long period of history, as the Bible and Quran denotes, has a little more complexity than the rudimentary concept of good and evil. And here the concept of the jinn is neglected by most researchers on spiritual and occult levels. Our jinn nature is that part of us not necessarily directly evil, yet extremely powerful. However, it is a part of us that even in mythology and psychology, has been neglected and misunderstood. We find ourselves, particularly in the West, steeped in a culture of what we can correctly identify as a heavily jinn culture. I am not talking about demonology or mysticism. The common understanding of jinn from Arabian folklore is a misconception, and distortion to say the least. The lore of the jinn as magical creatures with demonic powers, and not an archetypical aspect of our nature and inner psychology, has done a great disservice to understanding human nature, as well as distorting the meaning of the Prophet Muhammad's teachings.

The jinn nature is so important that the "Fathers in Heaven" based the whole matrix of their doctrine on it, and describe a monumental event in history so important as to apparently alter human nature itself, and we are relegating this awesome concept to a devil story! Indeed, the examination of our jinn nature along with a reexamination of our lower Adamic nature, and the conflict that our spiritual fathers are trying to get us to deal with, will lead to great insights into our selves that have been neglected.

The jinn nature is the transcendent aspect of being that deals with the extension of reality.

When we look at the jinn as some distant demonological or magical outer aspect of the universe, separate from ourselves, and not merely as a potential volatile part of our being, then we lose much in knowing ourselves. Much of the world today is the result of the activation or playing out of our jinn selves, particularly in relationship with our Adamic selves. These manifestations are not all new to us, just highly evolved compared to the past. The jinn self is responsible for many of the great scientific discoveries of the past few centuries. For it is our jinn nature, in the same way that the early Atlanteans and Lemurians had activated it, that is responsible for the huge sports industry, entertainment industry, medical industry, cosmetic industry, political industry, academic industry, educational industry - indeed the whole modern world in its huge mega, spiritual, religious, new-age manifestation. All of this is an explosion of our jinn nature, to a great degree. That in itself does not necessarily have to be viewed as negative to the welfare of the race, but of course, we know it could become a problem if we do not balance our jinn nature with consideration of our other natures. Balance and understanding is the key.

It is the jinn nature in us that strives for the unique feeling of the discovery of a scientific principle, or the exaltation of academic or sporting success, or the recognition of success by collogues in any field of endeavor. Certainly these examples can be countered by the observation that these simple, positive, ego-satisfying experiences can be carried to extreme levels, like megalomania and exaggerated egotism, as in the story in the garden when Iblis (feelings of superiority, excess egotism) of the jinn decided to undercut Adam because of these feelings. We all in life face similar situations when our jinn selves, which can be expressed as simple, minor, positive ego-satisfying experiences, as mentioned above, when allowed to run rampant and become too ego-centered ... well we all know what could happen: from the schoolyard bully to the scientist who distorts a medical experiment for fame, to the ego-maniacal dictator who oppresses his fellow man to maintain his out-of-control megalomania. Now the simple center of our lower Adamic selves, who primarily is concerned with safety and peace, and whose greatest enemy is fear and grief, is a counterbalance to the jinn nature going out of proportion. Indeed, when our jinn selves go too far with scientific experimentation (torturing animals, destroying the environment) over-stimulation with entertainment, excess of ego-centered activity, then the lower Adamic self can become frightened by the potential real or imagined havoc that the out-of-control jinn self can wreak on the lower Adam's peace. And there lies the conflict in a nutshell - for the lower Adam's center of operation, as well as safety and peace, being the primary center of our consciousness, overlaps with aspects of the jinn center. In other words, the lower Adamic self also has an aspect of it that craves psychic stimulation although it is not as high powered or as satisfying as the jinn stimulation centers. It is like the difference between looking out on a beautiful day and enjoying a lovely breeze while watching your child romp around in the grass - or hearing from a committee that you won a Nobel Peace Prize for chemistry.

These realities of our evolutionary nature are precisely reflected in the allegories in the famous *Book of Enoch*. This text chronicles the mixing of the jinn nature with human or lower Adamic nature. In *The Book of Enoch* there is a descriptive narrative about the punishment of the Samyaza and Azazyel inspired supermen, or as they are called, the watchers, because of their having sex with the lower Adamic females, thereby mixing the jinn nature with the lower Adamic nature.

As the Egyptian sage Hermes (or Idries to the Arabs) said ages ago, which stands as a standard of metaphysical doctrine: "What is above is as what is below." Or, in one sense, the outward reflects the inward. On the stage of the outward macrocosmic history, the story of the primordial Atlanteans and Lemurians, as well as the allegorical scriptural description of these events, there also is a simultaneous movement of the inner human. That would be the alchemical aspect of being in which the outward is but a manifestation of the inward. So we are seeing a journey of our own creation in understanding these allegories as they relate to our inner psychology.

On a more refined level of being, a level hidden from practically all history, movements in the inner soul - essence - of the human, which heretofore has never been spoken about, written about, by anyone anywhere on any level, is the most subtle aspect of our being. This reflects the reality of the fall of man, and does so on a level so subtle as not to be perceived by anyone until today, which we can describe as the creation of the *ellipse*.

Chapter 2

The Perfect Circle and the Ellipse

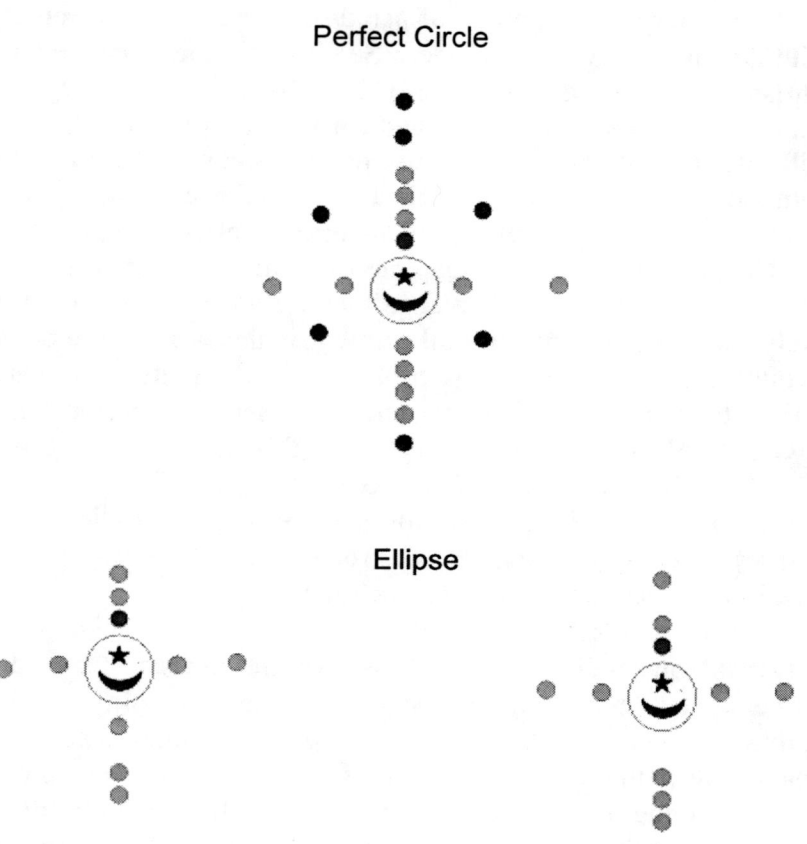

Figure 1

The human divine soul is a circle; it is a perfect circle. This term perfect circle only becomes necessary after we realize that in cosmic evolution, the perfect circle, or just circle, because of the fall of man, has become an ellipse - an imperfect circle. These are fitting metaphors for the soul - the quintessence of the human being. In recording this up-until-now unrecorded history of the human soul, we hope to fill in the blanks that heretofore recorded psycho-spiritual history has left empty.

In the story of the garden of paradise or the garden of felicity - with Adam and Eve - as Western Semitic religious lore, Judaism, Christianity, and Islam, has recorded, we have the basic allegory of not creation per se, but the fall and corruption of the human soul. I will refer to that corruption by use of the metaphor, *ellipse,* which etymologically means an imperfect circle: ellipse being a play on its opposite—perfect circle—the universal symbol of perfection..

The perfect circle is the Garden of Eden; it is where the history of all humans begins. It is a saga that we all are a part of, for it excludes none, and includes all of us. It is the story of the cosmic Evolutionary Intelligence (God) commanding itself (essence, spirit), to manifest itself to its uncreated self (the perfect man - HIGHER ADAM); then to its created self (fallen man - LOWER ADAM). This ladder of action is created in order that the EI can know an aspect of itself; or for the created self (fallen man - ellipse) to know creation and destruction. For he is the point of creation and destruction at the same time.

The perfect circle is a part our soul (inner makeup), the universal soul is God's essence. We are the microcosm. God, or all-being and non-being, is the macrocosm. In reality, there is nothing but God, but God is so huge we cannot measure or quantify him, that is why it is a sin of idolatry to designate any one thing as divine in exclusion to anything else. It is said that the sin which will not be forgiven, is idolatry, and the sin of idolatry is not saying this or that is God, but in excluding any other thing from being God, that is idolatry.

We are holograms or copies of God - that is, all of us, excluding NONE. Our being or consciousness is our soul, which is eternal and universal. The soul is the machine that drives our consciousness.

A part of that soul - in spiritual technical jargon - is called the *essence*. This essence is metaphorically called the Garden of Eden in the psychic-spiritual scriptures. A replica of this garden is the solar system, in which our sun is the center of a tree (lineage consciousness / extended circle) of a huge inter-dimensional being, whose lineage circle is the solar system.

In the metaphysical aphorism, relating to the soul: four worlds, seven levels, and infinite dimensions. This last part (infinite dimensions), is the part of the soul known as the *essence*. The essence IS THE GARDEN OF EDEN; it is the perfect circle that has evolved to an ellipse - the imperfect circle.

Therefore, it should be known that the story of the Garden of Eden is an allegory of the fall or corruption of that part of the soul known as the essence. This is apart from the other two aspects of the inner soul, the Holy Spirit, and the Tao. This we will cover in-depth later in the narrative.

STATES AND STATIONS

"And We said: O Adam! Dwell thou and thy wife in the Garden, and eat ye freely of the fruits) thereof where ye will ; but come not nigh this tree lest ye become wrongdoers."

Quran 2: 35

Here, Garden is symbolic of the source paradise or essence circle; while tree is the symbol for the essence states that I am here calling - metaphorically - circles (*Figure 1*). Fruits are the sublime, ecstatic states one gets when eating this celestial fruit, or invoking this consciousness inwardly.

Vine, tree, limbs (of the tree), lineage tree, (or) extension are all metaphors for the invocation - or entering into states and stations of consciousness. When a vine, tree, lineage extension, is invoked or entered into, a new tree - a physical replica of the previous tree - envelops the consciousness. Although the consciousness of the new tree is physically similar in terms of number of trees to the previous tree it came from, nonetheless it is a different consciousness-feeling from the previous tree.

In the beginning, the perfect man or Adam was owner of his inner Garden of Eden - paradise, as an uncreated reflection of the self of God (essence). He was master of the garden, he knew its laws and ways, and entered and enjoyed all the trees of the garden (states of pure consciousness) as an entertainment because he knew the science of states, and he had control of his states by way of control of his station.

In existence or being, the basic state of existence is the state made perfect and beautiful by acquisition of the station. The writers of the Bible and Quran use the symbol tree for these circles, or station-states of consciousness. In the perfect circle, all states and stations are perfect because they are uncreated, for it is only the created that has access to imperfection. The created can be perfect but only through evolutionary intelligence; therein is our hope and destiny. For we have been fallen by the hand of the EI; in order to know himself, and for us to know him. Even the perfect man "Our Fathers who art in Heaven" cannot know him like we can know him, for we, the lambs of God (the ellipse), the great sacrifice, are created. He said "I love *that I be known so I created man* (fallen man).

The perfect man knew the science of the self by birth and the science of the states by way of acquisition of the stations, or gardens wherein rivers flow - that is the garden of the soul.

These states were accessed by deep concentration, an art that was easy for primordial man (who was born with this ability), unlike today, where one has to practice years - sometimes decades - of concentration meditation to come close to the capacity primordial man had with his mind.

The inward science was very simple: there was a source garden (Essence) where nothing was before that and everything was after that, which was called merely the source, whose numerical value was the number five. It also was called the sun, also a circle. A replica of this inner and outer garden is the solar system. Each garden has at its core a powerful sun that serves as the center of the garden. As the planets revolve around the sun in our solar system, as the billions of others that we see and do not see in our universe, so the inner planets or (potential) states of our soul-consciousness revolve around our inner sun. The external is a replica of the real, or the inner, which is our real. The perfect being, Adam (higher and lower) in this garden knew all the laws of circular travel (science of states) or states of paradise; they were endless and eternal.

For example:

At station five, the source station, when one rested at that station, one had access to the paradise of the pure uncreated essence of the state therein. If one desired to go to station number 7 (number 7 station from the source station 5) and invoke or ascend in that consciousness, then one would enter that paradise and enjoy that station state. (Now recall, all of this is inner mental traveling). Then one would be at station 7 from station 5. At station 7, one could then ascend in station-state 12 from the 7, for example, enjoying that paradise. (By paradise I mean the pure consciousness bliss that one enjoys at each separate station-state, they are all different in feeling.) Then one could invoke the 7 state from the 12 (a repeat of the earlier 7, though of a different nature because it is the 7 from the 12, not 7 from the 5) station. On and on there is no end to these various states of consciousness-feelings of bliss, and pleasure *(Figure 2, and 3.)* Each state has a different consciousness-feeling. For instance, at the 5-circle state, one feels the awe and bliss of divine unity; in the 7-circle state one might feel the awesome happiness of freedom, at the 12-circle state, one would feel the essence of supreme safety in power. These are only analogies, for there are truly no words that can describe this paradise, one can only experience it. Recall that invoking a station-state means that it envelops the consciousness of the heart, and mind.

> 5-Source-circle-below, invoking 7-circle. 7-Circle is depicted emanating in consciousness - at right. Note: 7-circle position is expanded, reflecting what circle it is. Next the lower circle is emanated from the 7-circle, as the traveler chooses this circle [12] to emanate in their consciousness, following the 7.

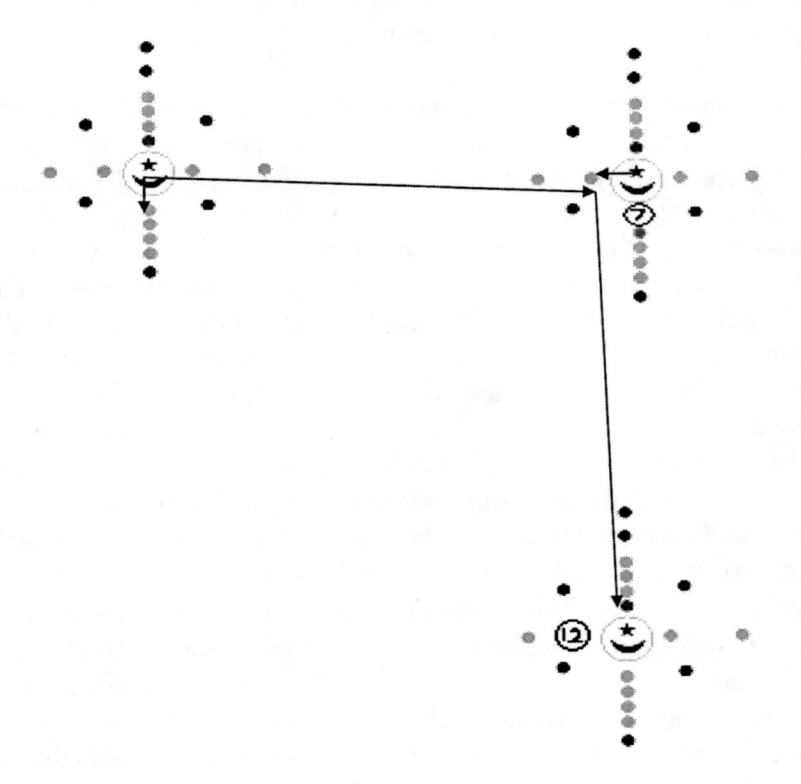

Figure 2: Lawful circular traveling: When one enters a station-state, the particular state circle becomes expanded [or colored red] this enables the traveler to determine what station-state they are in, at any time. The circle numbers are constant, at any state; the only change is in the red coloring or expansion of the state you are experiencing. So for instance, at circle 7, the 5-circle is still in the middle, but now it is a 5-circle of the 7- circle. Previously the 5-circle was the 5 of the 5.

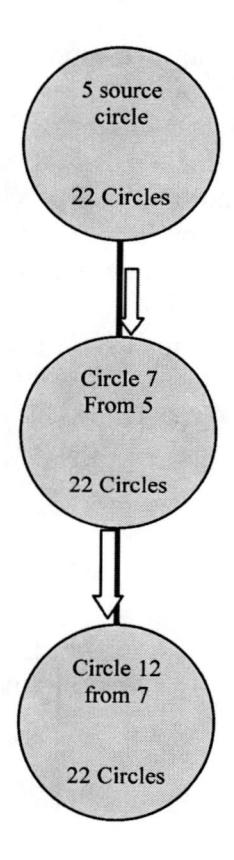

```
┌─────────────┐
│  5 source   │
│   circle    │
│             │
│  22 Circles │
└──────┬──────┘
       ▼
┌─────────────┐
│  Circle 7   │
│   From 5    │
│             │
│  22 Circles │
└──────┬──────┘
       ▼
┌─────────────┐
│  Circle 12  │
│   from 7    │
│             │
│  22 Circles │
└─────────────┘
```

Figure 3: Lawful circular travel

At each station (twenty-two circles), the core was surrounded by nineteen revolving spheres or stations. Two other spheres exist inside the core of any station. Therefore, there are twenty two circles in the entire essence, fifteen lower spheres, the core sphere (5) being one of the fifteen, and seven upper spheres, known as the angelic circles. There are twelve lower circles, and seven higher circles revolving around a core circle with two inner circles. A perfect state in which no state could be anything like the other or no two-station states could ever be the same.

The Ellipse: The Fall and Rise of the Human Soul

At each station-state (example above: 5-7-12) the consciousness or feeling of the person is different. They are blissful, joyful, though always different.

There are rules in circular travel and one rule is vitally important. The most important rule of circular motion is that at station 5, or the source station, *one could not go to circle 4 (Figure 4, and 5)*. (Shortly in the narrative will be displayed what happens when this forbidden tree is invoked.)

That circle is known *metaphorically* as the forbidden tree. One could only enter and operate in the 4-circle after one entered any other circle from the 5 - source circle OTHER THAN THE FOUR.

In other words, one could go from, for instance, source circle 5 to 7 to 9 then (4)
Or 5, 12 or (4)
Or 5, 13 (4)
Or 5, 2 (4)

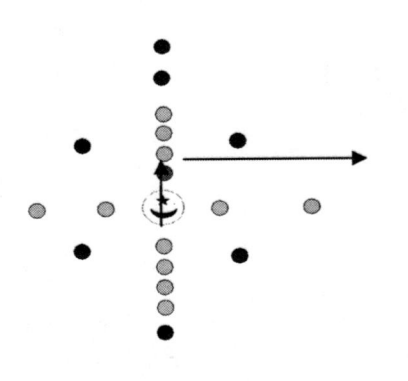

Invocation of (forbidden tree 4-circle) forbidden from the 5-source circle. The forbidden tree is colored a darker green to distinguish it from the others

Or 5, 8, 3 (4)
One could *not* do 5 to 4!

Figure 4: Unlawful circular travel. 5 to 4!

The Perfect Circle and the Ellipse

```
   ┌─────────┐
   │ 5 source│
   │  circle │
   └────┬────┘
        ▼
   ┌─────────┐
   │ 4 circle│
   │Forbidden│
   │   Tree  │
   └─────────┘
```

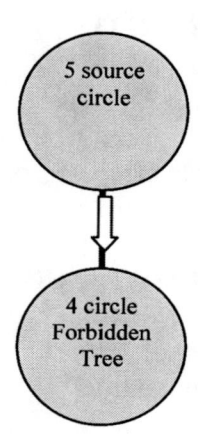

Figure 5: Unlawful circular travel. 5 to 4!

For the 4-circle, out of all of the circles, is a special circle, the circle of creation, and destruction, whose power and energy is too intense to enter at the source 5-circle.

In addition, if one entered circle 5 then 9, one could do any circle immediately next, but if one did the 9-circle that would lead back to the preceding circle, which in this case would be the source circle or 5. In other words, if one did 5, 8 then 8, one would return to the 5-circle. Alternatively, if one did 5, 9, 13, then 13, one would be at 9-circle.

Taking out, invoking, ascending in, entering into, drawing out, emanating, (eating fruit from the tree - the Quranic, and Biblical metaphor) all mean the same thing: inwardly, by deep concentration meditation, one leaves one station-state, and goes to another. The consciousness of a person becomes enveloped in that particular station-state of consciousness (fruit from the tree) for as long as they are there. They may return to the neutral 5-source station or descend further in another vine of consciousness, there is no end to this traveling. All these pure states of consciousness are indescribably blissful, that is why it is called paradise.

THE MEANING OF THE CIRCLES

Sphere 5: Source circle: Physical core point of celestial or soul circle, where all other circles revolve around. Includes, inside of it, circle 6 and 15. Also, vacant point of pure consciousness and timeless energy, the Lord, in Lord of the worlds. God principle.

THE LOWER CIRCLES

Sphere 1: The lower solar circle - represents the circle of theoretical truth, or truth in potential.

Sphere 2: The circle of the physicalization of truth - truth in form.

Sphere 3: The circle of imagination.

Sphere 4: The circle of nothingness, or the circle of creative potential and destruction. The circle known as the *forbidden tree*. The only circle in which an ellipse can be.

Sphere 6: The circle of the subtleties (lataifs): one of the 2 inner circles.

Sphere 7: The circle of the Lord of the earth. Concept of spirit presiding over form.

Sphere 8: The earth circle.

Sphere 9: The circle of binding.

Sphere 10: The circle of un-binding.

Sphere 11: The circle of the tariqa (inner).

Sphere 12: The circle of the sharia (outer).

Sphere 13: The solar circle, the circle of the sun.

Sphere 14: the lunar circle, the circle of the moon.

Sphere 15: The circle of purification or the circle of the chakras: one of the inner circles, also the dog (protective) circle of the lower circles.

THE UPPER OR ANGELIC CIRCLES

Sphere 16: The circle of Gabriel—circle of intelligence.

Sphere 17: The circle of Michael—circle of protection.

Sphere 18: The circle of Israfael—circle of expansion.

Sphere 19: The circle of: Azrael—circle of death.

Sphere 20: The circle of Kidder (earth angel).

Sphere 21: The circle of Malik—the King.

Sphere 22: The supreme Archangel Ashuara (the Dog-Star) the circle of evening (to even).

(See Figure 6 on next page)

Note: Number places on circle are listed in Annotations on page 185.

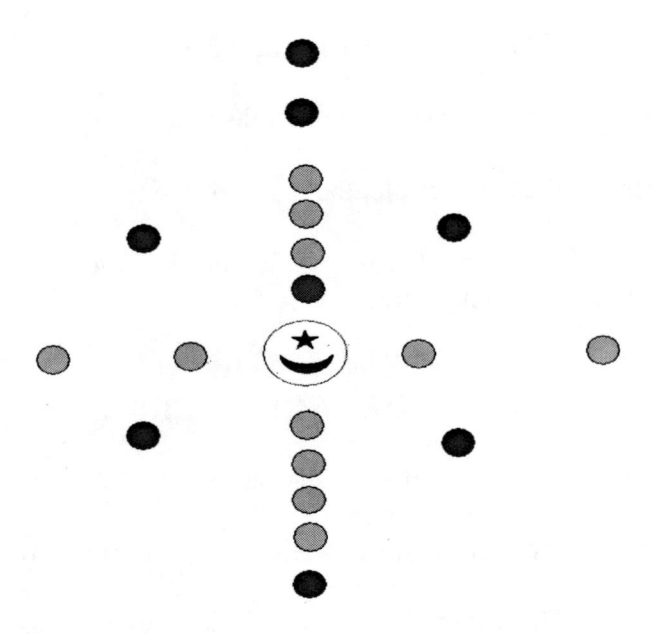

Figure 6: Illustration of perfect circle, Garden of Eden - paradise (essence)

THE SUBSTANCE OF A CIRCLE

The word circle is our metaphor for the term "tree of paradise" which itself is a metaphor for the essence. A circle, station-state is a condition of consciousness. When a state envelopes the heart by the person ingesting it or bringing it into the consciousness through the psychic spiritual exercise of concentration meditation (not the concentration meditation we know today which is not even remotely similar to the original), it is like taking a powerful drug that does not wear off. A circle is entered via a stargate of concentration that the consciousness enters and becomes overwhelmed by its state.

A circle, essence, or tree of paradise, is therefore essentially a state of consciousness. It was at one point in our history a stable state.

The Perfect Circle and the Ellipse

Human beings are *always* in a circle, or state of consciousness. We humans are presently at a low state of consciousness (unstable) because we have fallen from the original perfect states of consciousness we possessed before the fall. Therefore, many times - most times for many people - we are suffering, miserable often, and experience a state of overall dissatisfaction. That is why we designate this condition an "ellipse" a term that denotes an imperfect circle.

At the time of, or before the fall, we had control of these states and stations of consciousness, unlike today, where we go from various moods and various states of uncontrollable feelings, thoughts, ideas, and emotions that render this present condition generally miserable. Before the fall, no conditions existed such as those I mentioned above, for any human. The states of consciousness in any circle were stable, intensely pleasurable, exciting, and immensely rewarding and peaceful, and most important, balanced.

And (unto man): O Adam! Dwell thou and thy wife in the Garden and eat from whence ye will, but come not nigh this tree lest ye become wrong doer.

This is a reference to the fact that in our pre-fallen times we could ascend or descend in any state or station in the Garden of Eden through deep concentration. These inner states are deeply hidden in us, today they are virtually unknown to us, as opposed to primordial man who easily entered and returned from these lofty paradise states at will. Even the methodology of accessing these subtle sublime states and stations of bliss is long lost.

In our present elliptical condition, when we do experience any real glimpse of happiness, peace, stability, or joy, it is only a glimpse of the rewarding states we had in primordial times. Like a flashback in an acid trip, to use a profane analogy. This is true because we are in a corrupted or perverted circle or state that is a poor replica of its original condition. The power and intensity of these states is so enormous that we can't help feeling their intensity often, in glimpses, though nothing like it was in the

The Ellipse: The Fall and Rise of the Human Soul

period we enjoyed access to, and control of, them in primordial times. It is the negative feelings - fear, hatred, alienation - that are direct causes of the ellipse, or the corruption of our essence.

The original states have no degree of these feelings, which are energy-based - as all feelings are. Of course, these are negative energy-based feelings that will always exist as long as we are in the ellipse.

The apparent relative "stability" that we appear to experience is because we have consciousness of the three inner circles: 5, 6, and 15 *(Figure 7)*. The 6 and 15 hold the basic elements of our consciousness: the lower mind, the Lataifs of the Sufis that reside in the 6-circle *(Figure 8)* - that are light-based, and the Chakras of yoga in the 15-circle *(Figure 9)* - that are fire-based. These are not circular states, in other words, one cannot extend them like the other 19 circles. They are the preservers of our fundamental life, being, and consciousness.

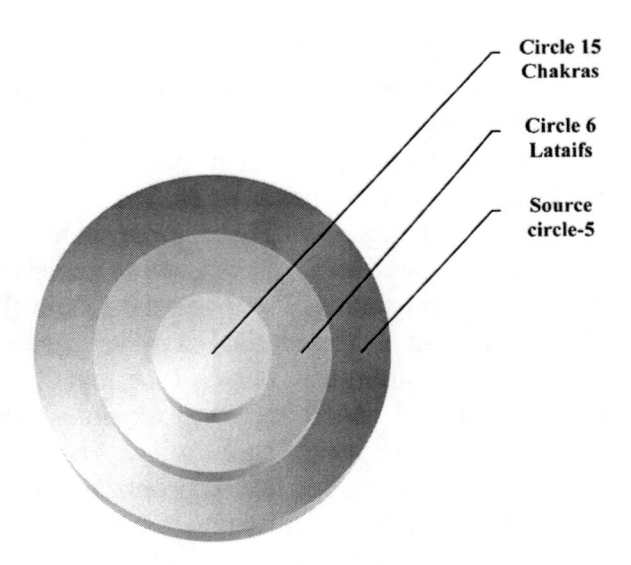

Circle 15
Chakras

Circle 6
Lataifs

Source
circle-5

Figure 7: The three inner circles are not circular states, in other words one cannot extend them like the other 19 circles. They are the preservers of our fundamental life, being, and consciousness.

The Perfect Circle and the Ellipse

LATAIF COLOR CODE AND BODY CORRESPONDENCES

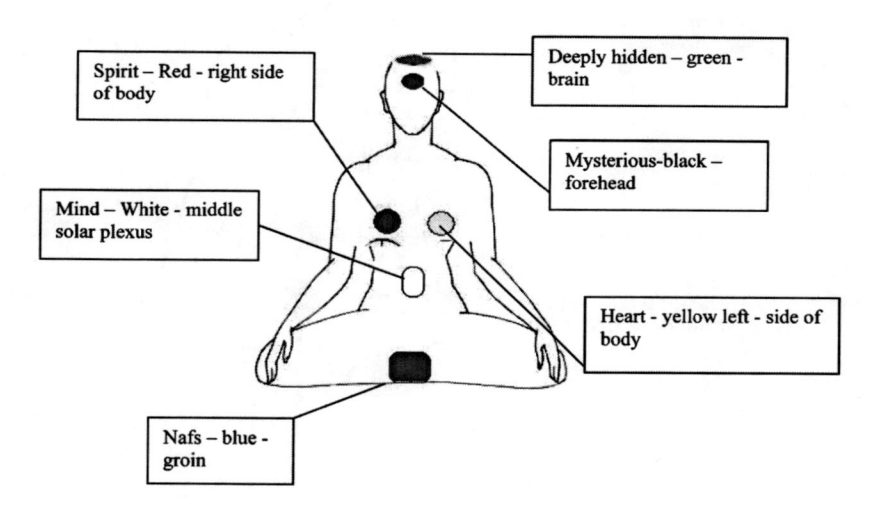

- Spirit – Red - right side of body
- Deeply hidden – green - brain
- Mysterious-black – forehead
- Mind – White - middle solar plexus
- Heart - yellow left - side of body
- Nafs – blue - groin

Figure 8: Lataifs are the subtle organs of perception that reside in the six circle. From the Sufi tradition, Lataifs are concentrated upon to gain enlightenment. These like the Chakras are associated with parts of the body, and color-coded, though the correspondence of the Lataifs with certain parts of the body is not literal, as are the Chakras.

CHAKRA CENTERS

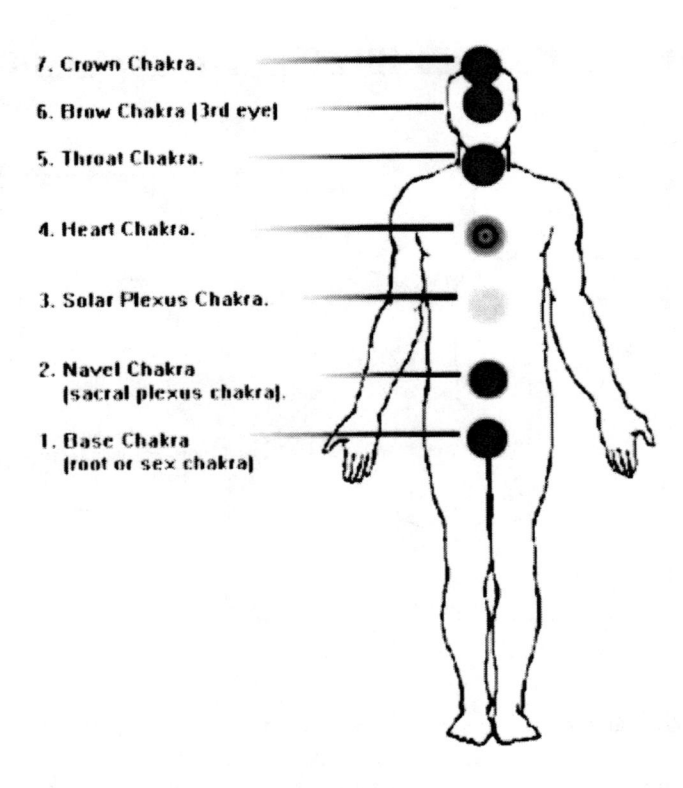

7. Crown Chakra.
6. Brow Chakra (3rd eye)
5. Throat Chakra.
4. Heart Chakra.
3. Solar Plexus Chakra.
2. Navel Chakra (sacral plexus chakra).
1. Base Chakra (root or sex chakra)

Figure 9: Chakras (Wheels) associated with circle 15 are from the Yogic tradition, and are energy vortices that are associated with the parts of the body shown above.

The Perfect Circle and the Ellipse

The 5-circle, the core or sun circle, the principle of pure radiation or nothingness, or unity, is all of our inner core being that all the other 19 circles circumnavigate when the circle is at its perfection, not as an ellipse as it exists now.

In terms of the physical aspect of a circle? Well the best idea of this is to look at one of the planets as a particular circle. The solar system is a direct reflection of the structure of the Garden of Eden, and the sun is the source station and the planets are the circular states and stations that the being could ascend, descend in, and animate that consciousness anytime they wish.

A circle is entered through stargate technology, which is the science of going in and out of states of consciousness-circles; it is a conduit from one circle-dimension or (world-stage-state) to another. The stargate also sends the individual back to the original circle, or preceding circle or circles if one wants to do that. This stargate sends the consciousness to the new circle where it binds the heart to it and the individual senses the new state and tastes its consciousness-feeling.

Much of this New Age, the movies and TV shows about stargates and wormholes, has come about because this psychic technology is slowly returning to humans' consciousness as we progress back to the ancient knowledge we lost. The brilliant new-age mythologist William Henry is a great source for knowledge of this lore.

These states are indescribably sublime, and each state is different from the others, and produces pleasurable feelings unlike anything humans could imagine. No drug or other artificial device can match this experience. In fact, drugs, alcohol, movies, sex, and things like this that humans have indulged in are only their unconscious attempt at garnishing some of the primordial paradise states we had before the fall. This, though, has to be reacquired through spiritual wisdom and science, these (unconscious) artificial attempts at duplicating this ancient spiritual technology through these means will not produce anything because this knowledge starts with controlling oneself through legitimate metaphysical sciences. Remember, this is all imprinted on our soul history, we have no choice but to practice artificial examples of what we did in primordial times in the paradise therein.

Chapter 3

The Story of the Circles

"As above, so below"
Hermes

Eons upon eons ago, the Evolutionary Intelligence manifested a microcosm of himself.

And so arose in it the six natures of being and the spirit.

Eons ago the universal soul entered circle 22, then 21, then 20, then 1, then 8, then 2, then 7 from upon the highest station of stations and then He entered the 4 station and emanated himself to manifest a hologram of himself, thereby manifesting the perfect or transcendent Adam. He taught Adam all the names and how to exist in the circles. Adam was not one being but many.

Eons ago the emanation of the transcendent Adam, the soul of souls, entered circle 21 then 1, then 8, then 4, then he projected the first part of the great jinn.

Eons ago, but later than the creation of the first part of the jinn, the greatest soul entered circle number 20, then 2, then 7, then 4, then he projected the second part of the great jinn.

Eons after that, the great soul entered circle 1, then 8, then 2, then 7, then 4; then he projected the emanation of the lower Adam of the kingdom into being.

The divine nature - the spirit of the universal soul - breathed in the holistic being all of these. The two aspects of divine nature are: The feminine, ESSENCE. The masculine, SPIRIT.

(The holistic being - man - would only become a semi-complete reality at the advent of the lower Adam. He was in fact the beginning of the completion of the holistic being or manifestation of God in the microcosm.)

All the manifest beings, the transcendent Adam, lower Adam, the angels, the jinn, and the animals reflected the grounding of the powerful energies (natures) that the EI breathed into the microcosmic soul. Each respective being reflected their essential natures, although not exclusive to the other natures. For instance: the angels - creatures of light - were not all light, as the jinn - creatures of fire - were not all fire, so the transcendent Adam - beings of heaven - were not all heaven, as the animals - creatures of anima - were not all anima. So, the lower Adam a - creature of earth - who held down the earth energy, was not all earthly in nature.

The Quran and Bible do not mention directly that there were a whole race of Adams, and that the transcendent Adam numbered many. The lower Adam came to being in history as an offspring of the transcendent Adam. Both were instructed by the EI in the inner science of states or circular movement.

Although one sprang from the other, the two Adams were not of the same nature. One was basically heavenly with its degree of earthly characteristics, and the other was primarily earthly in nature, although with its degree of heavenly characteristics.

The story in Semitic Abrahamic religious lore about the garden of paradise, in Islamic lore, or the garden of good and evil in the Old Testament is an allegory. An allegory is a story whose elements, people, places and events are wholly symbolic, for example:

The two traditions when they relate the story about the Garden of Eden (story element) are using the term - Garden of Eden - as a metaphor for the essence, an aspect of the soul.

The physical location of the Garden of Eden (that is in the outer world) is only important as it relates to the lesson it affords us in cosmology, in that the physical location and historicity of the story reflects reality playing itself out over the cosmic ladder of the four worlds.

The lowest of the four worlds, known as the world of the kingdom is the well-known world of apparent physical reality and history we all have been accustomed to.

When the universal Evolutionary Intelligence (God) manifests an action, it is always from the standpoint of the highest world, the fourth world, or the world of the sovereign power. That action will in time have a reflection in the next lower world, known as the world of the domination, or the third world. The initial divine action will appear as one thing in the fourth world and appear as another thing in the third world, even though in reality the action or apparent actions are in fact the same. As in the next descending world, the second world, or the world of the dominion or psychic powers, the divine action will have another appearance, and on down to the lowest or first world, the divine action will have a reflection in the world of the kingdom, as the nature of that world reflects reality.

Chapter 4

The Fall Continued

The commandment of the Evolutionary Intelligence to the angels and (jinn) as the Quran relates the story - to make obeisance to Adam - was a command relating to the lower Adam. No command had to be given to the already existing transcendent Adam or higher Adam. The angels and jinn were already submitting, and had become accustomed to the perfect Adam, no command there was necessary. For centuries, before the fall, these natures had evolved to a settling in and smooth operation of being as history reflects in the successful and powerful historical cultures of Lemuria and Atlantis. What is not generally known is that while these great jinn cultures were flourishing, the cosmic being known as transcendent Adam was existing alongside them.

Our Father "who art in Heaven" is in fact the transcendent Adam, who the metaphysical teacher known as Jesus spoke about getting in union with. He was in fact referring to the fallen man returning to his inner transcendent nature, which he had fallen from. Jesus was speaking from the standpoint (symbolically) of the fallen man. He was not referring to getting in union with "God" in that statement, but was talking about the mystical inner reunion with the inner perfect Adam, then on to "God," or completion.

In primordial times, when the creation of the lower Adam was initiated by the Evolutionary Intelligence and the command came down from the EI to the existing natures of the angelic, jinn, and transcendent Adam to submit to the new reality, all went along with this commandment for a time. The Quranic story tells us that a part the jinn, specifically a jinn intelligence called Iblis, did not go along with the EI intention. The term Iblis is derived from the Arabic word balasa, which means "he became distant or remote"; despairing on the mercy of God, and separating himself from the holistic order.

THE STORY OF THE GARDEN

The allegory and the reality

The two allegories of the fall of man that we are generally familiar with in the Quran and Bible say - in the case of the Quran - that the devil enticed Adam to eat of the forbidden tree that God told him not to eat of; this was the tree of immortality. And in the case of the Bible, it was the tree of good and evil. Two different allegorical references in fact to the fourth circle in the garden of states referred to above.

"120. But the Devil whispered to him, saying: O Adam! Shall I show thee the tree of immortality and power that wasteth not away?"

Quran 20: 120

"but of the tree of the knowledge of good and evil, thou shalt not eat of it; for in the day that thou eatest thereof thou shalt surely die."

Genesis 2: 16:17

Indeed the inner science of states that the transcendent Adam had been practicing for millennia, the new Adam was taught by his Father or Mother in "Heaven," the transcendent Adam, for a long period. The new Adam, the offspring of the perfect or transcendent Adam, numbered thousands over generations before "he" finally fell to eating of the forbidden tree, which was an offspring of the new (lower) Adam of some level of farther descent from the transcendent Adam. The fall was the spiritual novice, the new Adam entering the fourth circle from the 5 or source circle, which was forbidden by "God" and which all knew or was told would cause doom for he who eat thereof as the above verses denote.

Did the "devil" or in reality a group from the jinn cultures of Lemuria and Atlantis have anything to do with this fall as the allegories in the Quran and Bible state?

THE INNER CONDITION OF THE JINN

In the case of these particular jinn, their inner condition is of the essence of a four circle being. In other words a jinn is somewhat similar to the human in that he or she is a 4-circle "creation" of The EI; but unlike the human is NOT an ellipse. They are made of fire so they can balance that powerful element. Therefore, they are very powerful, intelligent, and have immense psychic capacities and energy in order to fulfill their mission in energy balancing in the cosmic order. In our story, the jinn who rebelled became aware of their cosmic function, and their relationship with the lower Adam and, as the Quran states, became dissatisfied with this.

"He said: What hindered thee that thou didst not fall prostrate when I bade thee? (Iblis) said: I am better than him. Thou createdst me of fire while him Thou didst create of mud."

Jinn as evolving sentient beings have a certain assigned period to do this balancing work before they are allowed to evolve to the transcendent Adam level. In the case of the Iblis line of the jinn, they have devolved backward because of their rebellion, similar to the devolution of man through his descent in the ellipse.

These powerful psychic adepts had little knowledge of the circular science (science of paradise) compared even to the new Adam, let alone the primordial Adam; they did nevertheless know that they (the jinn) themselves were 4-circle beings who were forbidden circular travel until their function of holding down the solar heat (FIRE) was concluded and then they could evolve to the level of the primordial Adam.

The Fall Continued

So from the two allegories we know that say the devil enticed Adam, the Quran is more specific in that it says that Adam did not deliberately disobey God, but forgot "the commandment of his Lord,"

"And verily we made a covenant of old with Adam, but he forgot, and we found in him no resolve to disobey"
Quran 20: 115

This is very important in that we know that the "devils' cosmic duty" was to hold down the energy or heat of the sun, thereby shielding its powerful energy from the new Adam. They had intuitively come to understand their own nature to some degree, and the nature of their cosmic function, and had become over time unhappy and envious of the Adamic races, upper and lower. So consequently what the jinn groups did to assist in the great fall of Adam can be deduced by the allegorical reference in scripture, in that it says "The devil made manifest to Adam their shame." This is a reference to the fact that the great jinn races, having intuitively come to understand their cosmic function of shielding Adam from the sun's heat, had by a powerful occult act or acts intervened in the lower Adam's consciousness to communicate to him - the lower Adam what he might be exposed to if it were not for the actions of the jinn in shielding Adam from the sun's heat. "Make manifest their shame." Or as the Bible says "taught him good and evil."

The intervention of the jinn or "devil" to mislead Adam was therefore not only a suggestion from the jinn to enter the fourth circle from the 5-source circle (something forbidden by God) but also the jinn telling Adam of his potential vulnerability in the jinn's cosmic protection of Adam, by shielding him from the sun's heat. This exposed Adam to the psychological state of shame and uncertainty, which in turn exposed him to the next machinations of our jinn's rebellion. That would be these jinn performing a powerful occult curse or spell on the now vulnerable Adam that caused a psychic fog so immense as to cause him to fall into a state of "forgetfulness" so intense as to allow him to forget the command of commands and enter the 4-circle from the 5 source circle. This spelled doom for Adam and his entire offspring, which would eventually produce as the Quran says "descend from this state," the ellipse or fallen man became a reality!

ADAM AFTER THE FALL

Two ellipses

This event in cosmic history is so immense as to literally cause a severing in the unity of the holistic being, the microcosmic expression of the macrocosmic God. That severing was the result of him entering the four circle from the 5-source circle. Eventually this produced an inward state so discombobulated as to become an abyss of imbalanced consciousness - resulting in the ellipse *or* imperfect circle, from the perfect circle *(Figure 10)*. This was in fact such a breach or severing of the newly fallen Adam from the transcendent Perfect Adam - who did not fall - as to render the holistic being, the God-man (Adam, upper and lower), as two separate beings, the one fallen Adam because of his "descent into the abyss" or ellipse, becoming a shadow replica of his past exploits, and a distant shadow of his father - transcendent Adam, or higher Adam.

The forming of the ellipse or fall of Adam was a trauma in the history of our "creation" and literally a tearing apart (although not permanent) of the inner soul which in fact could only be torn apart because the energy of the 4-circle from the source 5-circle is so powerful as to literally tear apart a bit of the soul of the fallen Adam and to change him into a new creation or a mutilation of his original being or nature.

The cosmic circular numerology of this cataclysmic event which resulted in the consequent history of our sojourn on this plane of existence after the fall is the following:

The lower Adam in the descent into the 4-circle from the 5

The Fall Continued

source circle split the fallen Adam into two different beings, with two separate elliptical states or conditions literally based on the ripping of the solar irradiations or lights that keep the soul aligned to the Holy Spirit; that the created ellipse became a solar system with two of its planets thrown out of the orbit of its inner revolution, in this case the following description of this tearing is such: The first ellipse saw the ejection of circle 1 and circle 8 from the inner solar orbit - that is in the first created ellipse, *(Figures 11, 12)*. Also in this sphere the higher angelic circle number 21 began to circumnavigate around circle number 22. However, this is not depicted in the graphic.

The second created ellipse saw from the solar sphere the 2, and the 7, circles ripped out of orbit as in the first created elliptical man *(Figures 13, 14)*. Similarly, the 20th angelic circle began to circumnavigate around the highest angelic circle, number 22. In cosmic astrological history, this is reflected in the cosmology of the great Dog Star Sirius with its two great orbital stars b and c.

Also, the number of the soul went from the divine 22 to the human-elliptic 13: Because as the 2 lower circles in each new respective fallen being (that is the 1-8 ellipse, and the 2-7 ellipse) ejected from their orbits of rotation, consequently the lower Adam (by the cosmic law of cause and effect) lost access to his higher angelic circles, that numbered 7 *(Figures 12 and 14)*. Therefore, with the lost of his immediate consciousness of the 2 lower sets of circles in each respective ellipse being (the 1-8, and the 2-7), with the additional lost of consciousness of the seven angelic circles; the lower Adam went from a celestial number of 22 to 13. In other words, the very circuitry of his souls-essence was severed; no longer celestial but now, terrestrial, and very human!

PERFECT CIRCLE 22 CIRCLES

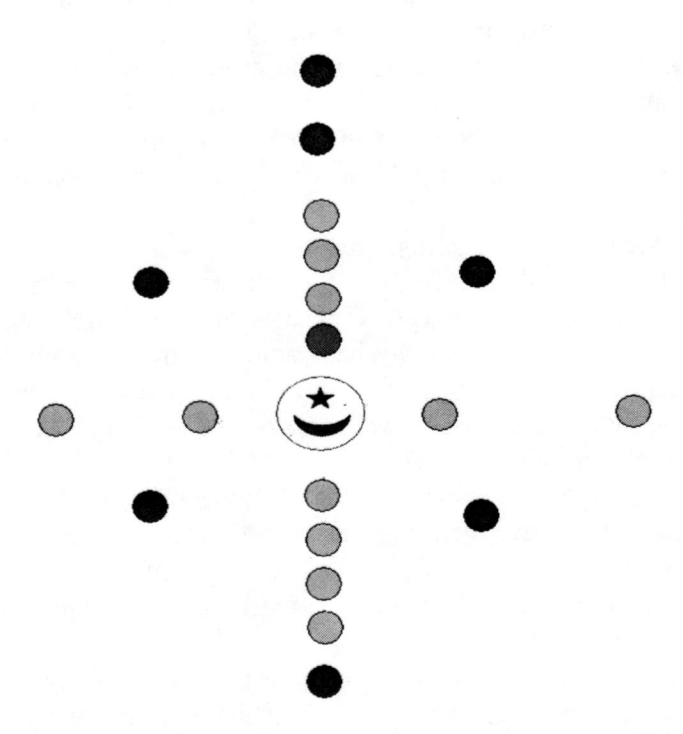

Figure 10: Illustration of the 22 circles of the essence in perfection.

The Fall Continued

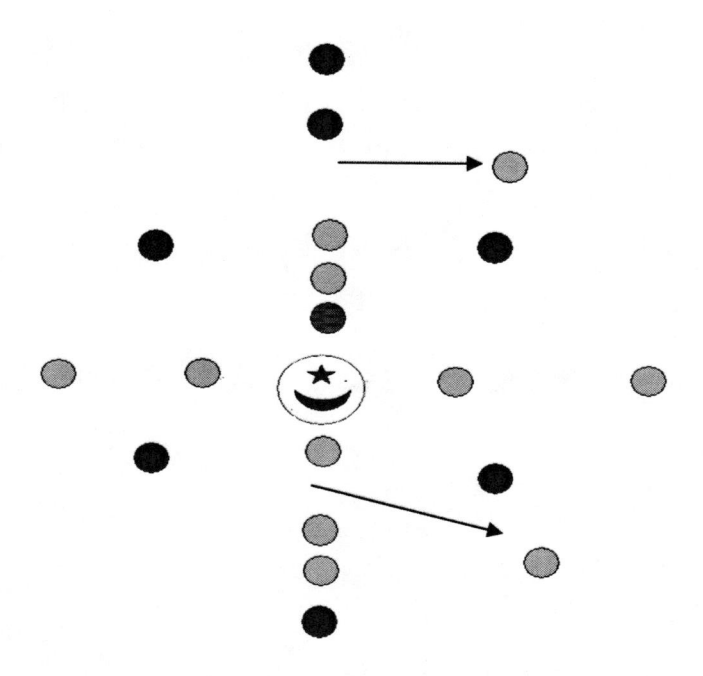

> **1, 8 ellipse**
> The descent in the 4-circle from 5-source circle begins to create the ellipse by ejecting 1 and 8 from 4-circle Garden of States. Underneath is view of 1, 8 created ellipse.

Figure 11: Illustration of ejection of circles 1, and 8 from essence sphere, or descent into 4-circle, forbidden tree.

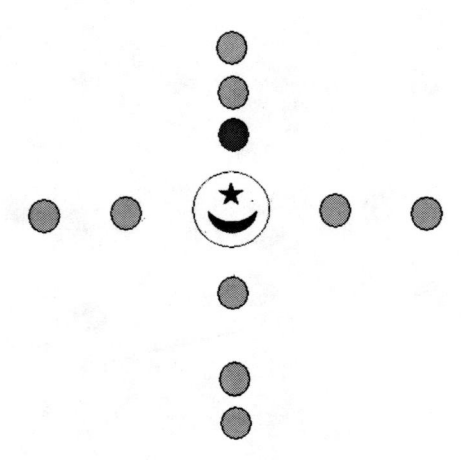

Figure 12: Created 1, 8 ellipse with 13 circles from the original 22 essences.
ELLIPSE-13 CIRCLES: THE CIRCLES AT POSITION 1 AND 8 ARE MISSING AND THE (7) ANGELIC CIRCLES ARE NOT SEEN.

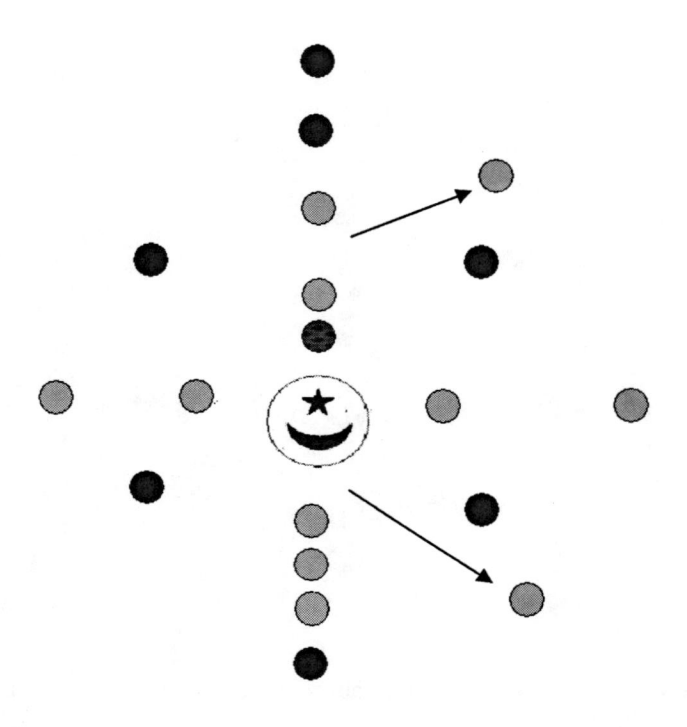

The Fall Continued

2, 7 ellipse
The descent in the 4-circle from 5-source circle begins to create the ellipse by ejecting 2 and 7 from 4-circle - Garden of States. Underneath is view of 2, 7 created ellipse.

Figure 13: Illustration of ejection of circles 2 and 7 from essence sphere, a descent into 4-circle.

The Ellipse: The Fall and Rise of the Human Soul

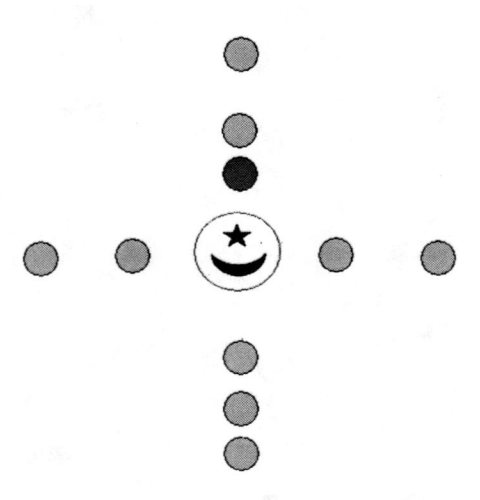

Figure 14: Created 2, 7 ellipse established with 13 circles from an original number of 22 essences.
 ELLIPSE-13 CIRCLES: THE CIRCLES AT POSITION 2 AND 7 ARE MISSING AND THE (7) ANGELIC CIRCLES ARE NOT SEEN.

The Fall Continued

THE FOURTH CIRCLE

The forbidden tree

"Your Lord has only forbidden you this tree lest you should become angels or immortals, and he swore to them saying: Surely, I am a sincere counselor unto you. So he beguiled them by deceit." (7:22)

A cosmic essence of God is found in the fourth circle, which is the circle of creative potential and the circle of destruction or nothingness. Herein the adept can harness great powers of creation as the verse implies, but remember the fall was not in going to this tree per se, but going to it from the wrong place. This powerful essence from the standpoint of the 5 - source circle we are now learning is fatally instructive in that we have as a race been suffering this lesson for millennia, and in fact only by the clock of the EI will we return to our cosmic perfect selves - transcendent Adam, then beyond that, to the state of completion, or expansion at resurrection as the Quran says: **"They will not desire removal therefrom."**

This world we live in and all its violence is a 4-circle ellipse. That is why it is so violent; this by the way does not include all the universe. It is only our solar system that is an ellipse, not the rest of the universe, although there certainly are other ellipses in the universe. The reason creation and existence is in this sphere of being is the power of the 4-circle. It is the circle the God-being uses to bring forth apparent creation with its rich beauty and awesome power as we witness in this world daily.

THE FIVE NEGATIVE ARCHETYPE ENERGIES THAT CAUSE EVIL

The powerful energies in the fourth circle from the five source circle (the forbidden tree syndrome) are reflected in the known archetypal "negative" energies described in the Bible and Quran, metaphorically referenced as *Gog and Magog*, whose lowest level energy manifestation produces violence, aggression, and the yin-yang opposing poles of restrictive and unrestrictive thinking. *The Dajjal* (anti-Christ) energy that produces the addiction to psychic-religious-matrices, that can produce religious deception, intolerance and bigotry. The *"Satanic"* energy that produces extreme perversions of natural inclinations. The *Iblis* energy that produces pride, ego, and despair (*separation*), the energy that on one level is the source of all these powerful energies that have plagued us since the fall in primordial times into this created abyss.

These "negative" energies are the result of the creation of the ellipse or the ellipse affect on the psychology of humans as elliptical. Over the years, this has manifested itself into what we experience as evil, and suffering. Of course, this brings to mind the verse in the Quran of the angels expressing their trepidation about the creation of Adam:

"And when your Lord said to the angels: I am about to appoint a vicegerent in the earth. They said: Will you place therein such as will cause disorder in it or shed blood ? We celebrate your praise and extol your holy names. He answered: I know what you know not." (2:30)

It is the ellipse that is creation, *(Figure 15)* and since we know that all "creation" is an alteration (or corruption) of natural existential reality, then we can understand the trepidation of the angels as recorded in the above verse.

CREATION VS EXISTENTIAL REALITY

> By existential reality, I mean God. Creation is an alternation (corruption) of existential reality in the sense that when the creative energy manifests itself it doesn't conjure up things from nothing (an impossibility even for God) but shapes and melds together from the 4-circle (the circle of creation, and destruction) substances that are already in existence. By mixing these materials in different configurations creates new kinds of reality. This is from existing materials, not magical conjurations from thin air. Nothing comes from nothing, as something never comes from nothing!
>
> Somewhat a prototype of the "forbidden tree" 4-circle is the Hindu mythological God Brahma, the god of creation. Similar to the 4-circle it has a destructive and transformative aspect in the god of destruction - Shiva, as one of its manifestations. As well as another partner in the preserver god - Vishnu, that fills out this mythological triad.

Figure 15: Brahma, the Hindu god of creation

Vicegerent is relative to the fact that the lower Adam's function in the holistic being is as a chooser of what is comfortable and acceptable to the organism in any situation.

NEGATIVE ENERGIES

Iblis: (devil) Pride, separation, isolation, vanity, excess ego, intriguer, and plotter. Extreme feelings of superiority, indicative of the devil in the classical understanding of the devil in the Garden of Eden.

Satanic: Extreme perversion of natural inclinations: Drug addiction, sexual perversions, violence, and deep psychological aberrations.

Dajjal (Anti-Christ): Religious fraud, self-deceived liar, hypocrite. Also, narrow minded dogmatist, literalist, and bigot. Personified in the ethnocentric bigot who uses and abuses race and religion to feel superior over others.

Gog: Stiff, hypocritically upright, uptight, vapid, narrow. The opposite of a smooth flowing nature.

Magog: The eventual result of the Gog energy that often-times produces war, violence, conquest, oppression, as our history records the recurring acts of conquest, plunder, murder, rape, and oppression.

The Gog and Magog syndrome work in a cruel yin/yang dance of extreme restriction to extreme unrestricted activity where the uptight Gog eventually, when all his posturing is used up, erupts in a storm of violence and expels the energy held in.

This ugly phenomenon has shown itself numerous times in spree and serial killings that particularly take place in America. This in addition to the traditional wars that humans always fight as well as ordinary violence that continually erupts in the experience of this world.

Mixed with the created ellipse, the root energies that exist in the 4-circle produce these negative archetypes that constantly plague humankind in a recurring series of evil, violence, injustice, and suffering, personified in wars of conquest, plunder, crime, and oppression, as well as myriad social and psychological pathologies that these energies perpetuate.

These energies are not per se negative as primal spirits in the 4-circle or essence, only extremely powerful from the standpoint of the 5-source circle. For these energy's (in the 4-circle) primary function under non-elliptical conditions serves as a ground point for all other sojourns of the God-being in 4 circular descents down the circular vines of consciousness-being within the essence circle. (This circular travel though is lawful and does not have the negativity we evolved to for going in the powerful 4-circle from the 5-source circle.) And this is the world that we are living in! It is as if we are living in a furnace that fuels a series of ovens that maintain the evils of our world.

With these powerful energies loose in our souls and in addition to the fact that we are - to use the metaphor of the mystic Gurdjieff - broken "machines" due to the *descent of our souls to an ellipse,* we have a double burden of dealing with powerful energies mentioned above while having to be sort of debased - I hasten to say - mutilated souls in that only 13 of our divine elements in our essence are even vaguely perceived or felt by us and it is as if we are a solar system not even spinning in orbit; or we are like a car that can never start up. Indeed man's horrible history of violence and evil, all patterned on the above stated powerful forces we have been exposed to in our fall to an ellipse, is like an ever-recurring nightmare which will never end. And we know it will never end, until the cosmic intention of the EI is fulfilled and played out in our sacrificial experience in the ellipse.

The negative energy generated in our souls is produced by the ellipse-perverted structure that we evolved to because we fell into these energies, that are only real because the perverted ellipse structure mixed with the 4-circle primal energy produces the output in our realm of existence. In fact, the commandment not to enter the fourth circle from the 5-source circle was because the energy is too ripe, pure and powerful for the consciousness of man to deal with from the position of the 5-circle. The negative experience of the human race since the fall has borne this out. That's why the four circle is to be entered only after one proceeds first down one of the other 19 circles (excluding the three inner circles that are not circular states) other than the four then it is "lawful" to enter the fourth circle which is for the purpose of the God-self to create and experiment with being, energies and essences for his entertainment and development. When the human species is done with our sojourn in the ellipse and has fulfilled our cosmic obligation, or fulfilled our punishment, some may view it as, we will return to our heritage as God-beings learning great lessons in the circular travel or the science of states, which is the natural state of the God being.

FINAL NOTE ON THIS MATTER:

It is vital to understand that because of the fall and the consequent events that included the lower Adamic beings essentially mixing our heritage with the jinn races of Iblis, we have inherited a great aspect of their work (the work originally assigned to the jinn) in shielding ourselves from the cosmic solar heat, as is intended for the jinn to do, not us. Consequently, this is another reason why we suffer immense violence, evil, ignorance, pain, and suffering in this world. In other words, we are now shielding ourselves from the sun's heat, no longer the jinn, who were equipped to deal with powerful energy, that we are not equipped to deal with.

Chapter 5

The Cosmic Intention of the Evolutionary Intelligence (God)

I mentioned above that the EI over time revealed to the jinn the reality of one of the primary reasons of their being: that is to balance the high-powered energy of the solar sphere in the universal energy grid *(Figure 16)*. This was fire energy, a reflection of their nature. As the Quran says, *"We created the Jinn of smokeless fire."* In other words, strange as it may sound to some, it was not only a rebellion in heaven, or mistake, or sin on our lamb, the lower Adam, that was responsible for the fall in the abyss of the created ellipse. It appears the very intention of the EI on one level, was for all this to happen (or that he knew its probability, and probably would happen) in order to create massive elliptical beings (that is most of humankind). The reason for this was to fulfill or play out a cosmic intention of his (God's), of destroying or TRANSMUTING elements of another world or dimension from (his) own inner solar 4-circle, in which he (God) no longer wishes to exist. This on one level is a lesson to us, for when humans return to perfection, then on to divine completion, we will understand creation and destruction; or, more precisely, transmutation of energy. We are in fact primarily hosts for carrying the energy from this 4-circle world to its destruction (I use the word destruction here almost as a metaphor, since nothing is ever destroyed, but only transmuted to another form of energy).

The Ellipse: The Fall and Rise of the Human Soul

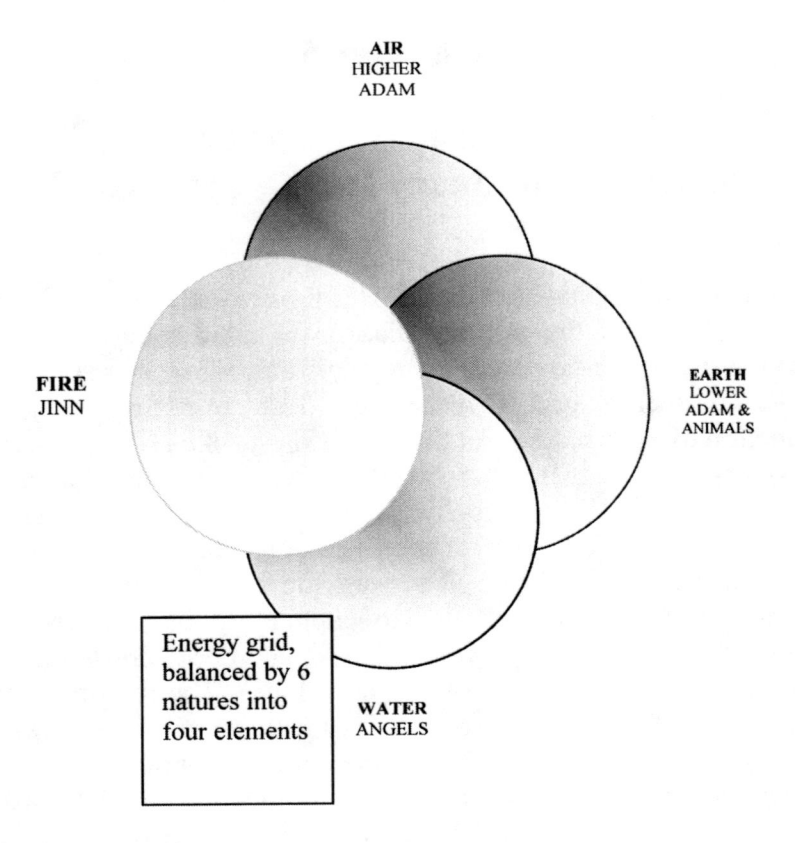

AIR
HIGHER
ADAM

FIRE
JINN

EARTH
LOWER
ADAM &
ANIMALS

Energy grid, balanced by 6 natures into four elements

WATER
ANGELS

Figure: 16: Symbolic illustration of universal solar energy grid made up of air, fire, water, and earth.

This other 4-circle world is not an ellipse, as our world and we are. It is the world that we as ellipses are projected to abide in by the EI precisely to destroy it! It is our negative charge as ellipses that has caused huge problems for the positive (non-elliptical) 4-circle beings. We have interacted openly and secretly with these beings throughout our history. The details of this are difficult to accept. I will therefore not go into this any further for obvious reasons at this time.

Sages have always told us: we are only *projections of God's imagination.* As strange as that sounds it is the primary reason for the choice of the EI for our sacrifice into the elliptical experience. In other words, we are the force of God's projected mind, cleaning out something for him!

This will be difficult to fathom or accept for many; that is, God is manipulating us in this pain cycle for a process of energy transformation, relative to his own self, at our expense and suffering *(Figure 17)*. But remember it did not have to happen this way. This energy transformation could have been done smoothly, as it was originally meant to be, but that would have taken much longer, and the price for this is our pain and suffering in this world's life. The benefit we will gain from this is superior knowledge of the four circle. Later in the book, we will cover this as relates to all of us being, as Jesus is understood to be, sacrifices for God's choices.

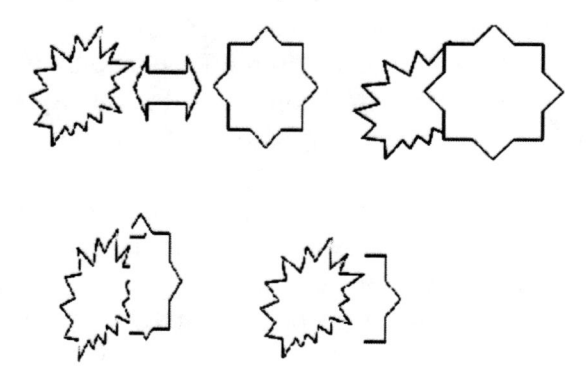

Figure 17: Top left, 4-world circle-ellipse (our world and self) created to transmute elements of non-ellipse 4-world. Our ellipse world exists within this somewhat latent 4-circle world. Illustrations on bottom right show our elliptical (negative charged world) clashing with the non-elliptical (positively charged) world in order to transmute elements of it.

Indeed, what we call, or have come to understand as God on one level, is the macrocosmic intelligent being whose essence, the solar system (ellipse), is one of his/her station states. We are a product and reflection of that greater being that has sacrificed and obligated us to fulfill this mission. The fragile attempts at "perfecting humankind" by all the mystics, sages, and religionists will not curtail this mission. The bulk of humankind will remain ellipses until the work of subtly destroying this 4-circle world is done. The few individuals who in this plane of experience wrestle themselves out of the ellipse, in a return to perfection, are only reflecting the reality of the primordial Adam in "heaven" who did not fall. There were only thirty of them. In addition, this includes the few lower Adams who remained steadfast in not falling in the ellipse at the time of the great temptation. This archetypical inner space is our roadmap of return to God. That is the primordial or higher Adam and the few lower Adamic beings that did not fall. Consequently, in our psychic history, as reflects those thirty primordial perfect beings (of whom many have long evolved to the higher state of completion) a very few people will beat the imperfect circle. This is reflected in the Prophets, Saints, Buddhas, and their few sincere successful followers that history records may have escaped the ellipse. I say may, because in the mystical sciences one can acquire a degree of "enlightenment" even as an ellipse based on the methodology of centering one's consciousness on the center of the primary core circle(s) 5, 6, and 15, that although are part of the ellipse, are nonetheless still very stable. This is, though, certainly no replacement for the time when we complete our obligation that the EI has imposed on us, and are allowed to return to primordial perfection, then to completion. As the Quran reflects this period in the Surah Al Asr:

BY THE TIME,
SURELY MAN IS IN LOST
EXCEPT THOSE WHO BELIEVE
AND EXHORT ONE ANOTHER TO TRUTH
AND EXHORT ONE ANOTHER TO PATIENCE

The EI uses our very imperfections, caused by the creation of the ellipse, which as we know in experiencing this difficult world, inclines us to negativity by the very nature of the imperfect circle. That is a deliberate design by God, it is not an accident, albeit not his original intention.

Suffice it to conclude that:

How can a God, who has power over all things, allow something to get so out of his control, and that not be by HIS OWN DESIGN! THAT IS NOT THE NATURE OF GOD.

HOW CAN A GOD, WHO HAS POWER OVER ALL THINGS, ALLOW SOME KIND OF SO-CALLED DEVIL TO DO SOMETHING OTHER THAN THIS ALL POWERFUL BEING WANTS?

HOW CAN THAT POSSIBLY BE? WHERE DOES THIS DEVIL GET HIS INDEPENDENT WILL TO OPPOSE GOD FROM? ANOTHER GOD OR ANOTHER INDEPENDENT WILL SOMEWHERE?

BUT HOW COULD THAT BE? HE (GOD) HAS POWER OVER ALL THINGS.

HE HAD TO, AT ONE POINT, CREATE THAT VERY THING THAT IS SUPPOSED TO BE OPPOSED TO HIM, RIGHT!

WHAT INDEPENDENT WILL CAN ANYTHING IN BEING REALLY DO, OPPOSED TO SOMETHING THAT HAS POWER OVER ALL THINGS?

IF THAT WERE THE CASE THEN GOD DOES NOT HAVE POWER OVER ALL THINGS.

THERE IS ANOTHER INDEPENDENT WILL DOING SOMETHING SOMEWHERE ELSE.

AND WE KNOW THAT IF GOD IS ONE, AND HAS POWER OVER ALL THINGS, AS WESTERN SPIRITUAL LORE CONTENDS, THEN THERE CAN BE NOTHING OCCURRING OPPOSED TO THAT PRINCIPLE. ULTIMATELY ANYTHING ELSE EQUALS CHAOS OR DELUSION.

In the allegories in the Western Semitic Abrahamic tradition, it is always baffling when those scriptures seemed to speak about destruction of the world, and the perfection of it at the same time. If the world is to become perfect then why destroy it? Or why destroy a world you can make perfect? The reality that we are a world within another world serving as the host for the destruction of one of those worlds, and at the same time serving as the vehicle for the perfection of one of these worlds, is an answer to that above-mentioned mystery. The *good news* is that the completion of the work of the ellipse is close, and shortly the blessings and challenges of the great period of cosmic change will be upon us.

Chapter 6

The Spiritual Hierarchy

God does not want total chaos in this world, even though one has to conclude that he does want controlled chaos. This reality viewed in the best light is an explanation of the revelation of the religions of the fallen man whose very nature as an ellipse makes it impossible for him to benefit from religious instruction, unless he is willing to try to upend his created (secondary) nature - an almost impossible task. This is illustrated by the allegory in the Quran about the sojourn of the spiritual adept Dhul Quarnain in the land where Gog and Magog (two negative energies the ellipse generates, spoken about above) were doing mischief in the land:

Some Islamic theologians have identified the Macedonian conqueror Alexander the Great, or the Persian ruler Darius as Zul-Quarnain, as the person described in the following scripture.

18.83: They ask thee concerning Zul-qarnain. Say, "I will rehearse to you something of his story."

18:84 Verily We established his power on earth, and We gave him the ways and the means to all ends.

18.85: One (such) way he followed,

18.86: Until, when he reached the setting of the sun, he found it set in a spring of murky water: Near it he found a People: We said: "O Zul-qarnain! (thou hast authority,) either to punish them, or to treat them with kindness."

18.87: He said: "Whoever doth wrong, him shall we punish; then shall he be sent back to his Lord; and He will punish him with a punishment unheard-of (before).

18.88: "But whoever believes, and works righteousness,- he shall have a goodly reward, and easy will be his task as We order it by our Command."

18.89: Then followed he (another) way,

18.90: Until, when he came to the rising of the sun, he found it rising on a people for whom We had provided no covering protection against the sun.

18.91: (He left them) as they were: We completely understood what was before him.

18.92: Then followed he (another) way,

18.93: Until, when he reached (a tract) between two mountains, he found, beneath them, a people who scarcely understood a word.

18:94: They said: "O Zul-qarnain! the Gog and Magog do great mischief on earth: shall we then render thee tribute in order that thou mightest erect a barrier between us and them?

18.95: He said: "(The power) in which my Lord has established me is better (than tribute): Help me therefore with strength (and labour): I will erect a strong barrier between you and them:

18.96: "Bring me blocks of iron." At length, when he had filled up the space between the two steep mountain-sides, He said, "Blow (with your bellows)" Then, when he had made it (red) as fire, he said: "Bring me, that I may pour over it, molten lead."

The Spiritual Hierarchy

18.97: Thus were they made powerless to scale it or to dig through it.

18:98: He said: "This is a mercy from my Lord: But when the promise of my Lord comes to pass, He will make it into dust; and the promise of my Lord is true."

18.99: On that day We shall leave them to surge like waves on one another: the trumpet will be blown, and We shall collect them all together.

18.100: And We shall present Hell that day for Unbelievers to see, all spread out,-

18.101: (Unbelievers) whose eyes had been under a veil from remembrance of Me, and who had been unable even to hear.

The story is referring to the spiritual adept Zul Quran as a member of the spiritual hierarchy, who went about the land helping a tribe of people whom the Quran relates were having trouble with Gog and Magog doing mischief in the land. This story is a prime example of how the hierarchy sometimes intervenes in the affairs of the fallen man to keep things under control. But we see here the adept offering the people a path out of the ellipse when he says to them "That wherein my Lord has established me is better." The expression by the adept that he will raise a wall to keep Gog and Magog out for a time only until the wall crumbles, and then Gog and Magog will then be let loose again is an expression to the people that until they alter their nature and "get out of the ellipse" they will always, one way or the other, be subject to the negative energy form Gog and Magog.

This is an example of the conflict of God who on the one hand, perhaps we can say his left hand, making us ellipses to perform a cosmic function as sacrifices, and his right hand as a merciful God offering us a way out of the ellipse that God himself knows that he himself made practically impossible for us to get out of!!!! CATCH 22!!!! One can say, though, that his mercy is reflected in the annals of cosmic time as he has assured us that when our elliptical function is up, then we will return to our original state of paradise and beyond to a better resting place, wherein we will "not desire removal therefrom."

This spiritual hierarchy includes a triad of very advanced adepts that is reflected in the Buddhist initiation ceremony, and the chapter in Quran, Al Nas, or The Men.

Buddhists take refuge in initiation:
In the Buddha
The Dhamma
The Sangha

In the Quranic chapter, The Men, are these verses that are relative to the Buddhist ceremony:
I seek the refuge of God
The Lord of men
The King of men
The God of men

The Lord or men are the realized adepts who have come out of a mild form of the ellipse all the way to total freedom. These are essentially of the lineage of the lower Adam who did not fall. The prophets are all inclusive of this group. The first line in the Islamic book "Lord of men" corresponds to the first line in the Buddhist initiation "Seek refuge of Buddha."

"The King of men" are a group of adepts of the transcendent Adam lineage who have never experienced the ellipse and are destined to evolve to the higher stage of expansion soon, but as they are still at the stage of perfection, are a part of the hierarchy that rules us. This corresponds to "Seeking refuge in the Dhamma" (universal law).

"The God of Men" are the Goddesses such as the supreme feminine essences like The Virgin Mary, Isis, Kali, Diana, Tara - a direct reflection of the essence and other transcendent female essences that like the transcendent Adam have never experienced the elliptical condition. It is these "Goddesses" that by cosmic custom can only directly reflect the light of the pure essences. This essence relates to the Sangha in the sense that we are ALWAYS in a station-state as conscious beings, and the Sangha as community is representative of this from the Buddhist standpoint. We are ALWAYS in a community of station-states - essences.

This group of advanced adepts is the closest thing we will ever get to any anthropomorphic "God" of any sort. The entire spirit in all of them equals our God.

They will intervene at the closeness to destruction that humans many times come to, that is before the inevitable dissolution of the ellipse and in turn the return of the huge organism of this solar system to its source circle.

Remember, the world we live in is but a huge 4-circle of an organism that is using us to manipulate the energy in his 4-circle! The poem at the end of the book will clarify this.

It is the scriptural allegories in which the elliptical interpretation of history is to some degree derived from. There is no reality of God hiding what he has done. It is that verse in the Quran that records the angels saying to God as relates to the creation of the lower Adam: "Will thou create him to shed bloodshed and spread mischief?" - referring to the angels intervening in our history to basically cloak the reality that God himself "created," "evil" (by way of the ellipse) that is cloaked in metaphors and allegories that the Semitic spiritual masters (including the "angels") decided to hide from us until we are mature enough to understand this phenomenon. Consequently these fairy tales (literal interpretation of allegories) that the traditional religions have perpetuated on the world through ignorance and avarice by the traditional clerics, and caution by the spiritual masters and angels as the above Quranic verse implies. It was not the choice of the EI to hide from us what he did in sacrificing humans in the ellipse. In fact the very well known myth of Jesus is a testimony to this hidden reality. The sacrifice myth of Jesus is an *allegory* of the fallen man or Adam being sacrificed in the ellipse by God. In other words Jesus in fact was not a fallen man, but played out the myth as a representative of the fallen Adam. In other words it is not Jesus who is the *lamb*, but you and I, the ordinary poor human being, who is the true Lamb of God. For Jesus, as are all Prophets and Saints are replicas or *sons* of the primordial (transcendent Adam) or perfect man who is not an ellipse. These great men and woman who Buddha described as "having a little dust in their eyes" are not essentially ellipses, including Muhammad, Moses, Elijah, Chrisna, and all the holy men, prophets and saints of high rank, are not ellipses. This is of course a reality they (our spiritual "Fathers in Heaven" and angels) really don't want to emphasize to the general public, for they wish to maintain a doctrine in which we all have access to God's mercy and can reach the "truth" in time even though reality refutes this notion on the level of the ellipse period. It is only after the ellipse period is up that the playing field will be even.

This is not a criticism of our spiritual fathers, the Prophets, Buddhas, Saints, Avatars, Angels who indeed love us and whose intentions is noble beyond question, but the truth must be told of why their efforts have failed to bear fruit and that is because God intends for the imperfect circle to play itself out according to his intention to use our experience with negative energy (of which the ellipse is a host) to filter energy out of the mysterious 4-circle world of God in the macrocosm. That is our job here and it will be done until it is complete, and then we will as a race (some of us later than others, because they have transgressed on the rights of others) will return to perfection and beyond.

As stated above our "Father in Heaven" or, as the Sufis call the Qutb or spiritual center of the world and his circle, or as the scripture describes as "the angels" or some call the demiurge or spiritual hierarchy that "rules" for God in his "absence," in essence have never and will never be able to lead us to spiritual perfection, all they can do is visibly and secretly do things to hold the fragile ellipse together heroically so it can be maintained in order for the primary reason for its existence to be played out. That is the processing of energy in this mysterious 4-circle world that belongs to God (macrocosm) and he has produced us and the strange and painful aspect of our experience in this world to fulfill this mission. In fact this (God) we have come to "know" in a sense is a being whose "soul-essence" is the solar system.

This is not saying that the solar system is God, but that it is a part of God: (*soul*ar system) or system of the soul. Recall, this aspect of the soul is the *essence*. Later we will cover all the elements of the soul sphere and how in perfection it operates as a smooth running energy-machine of awesome capacity!

Then the traditional spiritual hierarchy or Demiurge's primary function is not to lead us to God but to maintain the fragile ellipse so that the primary function of we fallen beings can be fulfilled. That is not to say that the spiritual hierarchy in any way functions as a body to foster negativity; on the contrary, their job is to save us when the elliptical race goes too far in the matrix of this corrupt world - and reaches as it has many times the precipice of destruction. They do not advise us to do the negative things we do to each other; in fact as the Quran reminds us in the verse which refers to the angels speaking about the new Adam, they say: "Will thou create him to spread mischief and spread bloodshed." Consequently, the angels and the primordial perfect Adams - our Fathers in Heaven - inspired the religions to assist in maintaining some kind of attempt at morality to control the elliptical fallen race. This assistance only had to be rendered to humanity after the fall of man; there was no reason for religions to exist before the descent into the ellipse. On the rare but almost impossible occasion that an ellipse was to climb out of the imperfect circle and return to the perfect circle and his inner transcendent self, then the opportunity and guidance is there in the mystical paths that exist for us. As the Quran records God saying to the fallen Adam: Chapter 2: 35-39 of the Yusuf Ali translation.

2.35
We said: "O Adam! dwell thou and thy wife in the Garden; and eat of the bountiful things therein as (where and when) ye will; but approach not this tree, or ye run into harm and transgression."

2.36
Then did Satan make them slip from the (garden), and get them out of the state (of felicity) in which they had been. We said: "Get ye down, all (ye people), with enmity between yourselves. On earth will be your dwelling-place and your means of livelihood - for a time."

That (state) in which they were; and We said: Get forth, some of you being the enemies of others, and there is for you in the earth an abode and a provision for a time.

2.37
Then learnt Adam from his Lord words of inspiration, and his Lord Turned towards him; for He is Oft-Returning, Most Merciful.

2.38
We said: "Get ye down all from here; and if, as is sure, there comes to you Guidance from me, whosoever follows My guidance, on them shall be no fear, nor shall they grieve.

("Get ye down" is the universal law that transformed the lower Adam into an ellipse over time)

2.39
"But those who reject Faith and belie Our Signs, they shall be companions of the Fire; they shall abide therein."

DISTORTION OF GUIDANCE

Unfortunately the fallen man that the spiritual paths have been created to help and control are the ones who eventually destroy and mutilate the spiritual paths as history records that all religions originally founded by the inspiration of one of the transcendent Adams or angelic nature, always become distorted and changed by the people they are designed to help!

This is one of the reasons why the EI does seem to change his religious science. This change is because humans always in a short time distort the original message from him/her and turn it into something else. Therefore an upgrade of knowledge must always be if any real freshness of the doctrine is to be maintained.

One reason for the revelation of Islam on this earth was for a counterbalance to the distortion of cosmic truth that formal Christianity was perpetuating. Among other reasons, one was that Muhammad was the legitimate heir of the Abrahamic dispensation or the seal of the prophets, the other side of Buddha, who was the seal of the Arian dispensation of earthly spiritual guidance given to us after the fall of man.

Because of racial and religious pride, the "People of the Book" (Christians and Jews) at the time of Muhammad would not accept him (expected, not surprising) so consequently the world has faced violence and division in the midst of the "People of the Book", i.e. the crusades, the Arab-Israeli conflict, the present day war vs. Islamist terrorism, etc.

A strange thing indeed, a family of people, Christians, Jews, and Muslims, who are of the same spiritual progeny of Abraham slaughtering each other in the name of the same God! Or more perverse is the claim of same Christians that the God of the Moslems is not the same God of the Christians! Now we see why Buddha avoided God.

Inevitably, this will occur, and it is precisely because of the ellipse, as a natural degenerative process that will always do this to anything that comes into its milieu: that is distorting it, twisting it, until it no longer resembles its original condition.

For instance, the distortion of the master/teacher principle by the Christians is an established fact: making Jesus a God figure rather than simply a transcendent teacher or literally believing that he was the son of God. Similarly the distortion by the Jews of the nation principle in their chosen people doctrine. Now we come to the Muslim brothers, and what have they distorted?

The Ellipse: The Fall and Rise of the Human Soul

We may be witnessing a birth of the distortion. That being the jihad principle that for years in the Sufi doctrine has been recognized as a principle of inner struggle.

As the tradition has it that Muhammad spoke to the populace on returning to the camp after a battle with his enemies: "Now we have left the minor jihad and come to the major jihad."

If this religion is not to become enveloped by distortion as Christianity and Judaism has, then Muslims must be vigilant to stop this. I am not optimistic. It is ironic though to see the enemies of religion in fact encouraging another one of our major faiths going the way of ignorance, pretending to be against this. It may be that the same occult groups that distorted our two previous faiths are at work again.

The religions are telling us some truth in their doctrine of slavery to God, which reflects the reality of this affair. We are and have been God's slaves, even though this concept to many modern spirituals is anathema to their New Age ideas. That's fine, and only reflects us evolving slowly out of the slavery of the ellipse as the reason for its being fades into cosmic insignificance and we can evolve to the heights of our spiritual destiny. God has indeed hoisted on us a great mission of supreme sacrifice, the ordinary man and woman. We have nothing to be ashamed of, but only those of us who have taken advantage of the situation and do injustice to their fellow human being, they as the lore speaks will unfortunately by the laws of what has come to be known as karma will encompass them, then the spiritual realities in the last days will appear to purify the ellipse and the steadfast or righteous ones will be separated from those who have done injustice to others by their own free will (they were not compelled or advised to do this injustice to their fellow humans, the reality of the ellipse will not be an excuse for them); they will enter what is known in religious terminology as *hell*.

Indeed this hell period, graphically depicted allegorically in the Quran is *on this earth*. It is not any fiery place underneath the ground as the paradise is also on this earth and is not any ethereal place in the sky. Hell is and will be an outward as well as inward reality in the coming age of purification. The numerology of hell is clearly depicted in the Quran and confirms exactly the elliptical cosmology as we will cover later.

INVISIBLE HIERARCHIES

Universal laws

Due to the universal subtle laws of the cosmos (the unseen hierarchy) the macrocosm and microcosm are destined to return to their perfect nature and beyond. In fact to a greater state, that is completion, which is called resurrection (expansion) in our western religious jargon. This is based on three universal laws.

1. As above so below: Theoretically, if a solar system (station) evolves to an ellipse, for whatever reason, the microcosm has no choice but to evolve similarly. Ironically, the macrocosmic solar system in our affair evolved to an ellipse because the microcosm evolved to an ellipse first via the fall! As when the macrocosm re-evolves to a perfect circle, we as microcosm will have no choice but to evolve to a perfect circle. (The pain, suffering and duration of abiding in the ellipse depend on the energy we expend in getting out of it.)
2. Return to God: Every separated energy form - however that occurred - that is existential and transcendent (always been here, and always will be here) returns to its source. God the ultimate perfect circle is our source, but that is God in the macrocosm. Examples of microcosms returning to the source are Muhammad, Buddha, and Jesus. This indicates the good news that one day ALL, AND I MEAN ALL, WILL RETURN TO PERFECTION.
3. The law of the merging of heaven and earth: The closer we get to the return of the elements of the soul of the macrocosm to

their original position in the circle means that heaven and earth will merge. What does this mean? It means that time will accelerate. When time accelerates things in the world will begin to become very strange. For example, criminals will be punished rapidly. Justice will become swift, and other things will occur I cannot mention!

Parenthetically, another important phenomenon is the universal law of expansion (resurrection/union) which I will not attempt to delineate in depth here, but in later chapters.

It is the ellipse (only) where all this will take place, in the macrocosm and microcosms. The ellipse relates to *this* solar system. This solar system is a branch of the Garden of Eden-*Essence* (perfect circle) a part or extension of the greater galaxy.

Ellipses are resolved all the time, when an individual microcosm reaches what is called enlightenment, awakening, or completion. There are not many who succeed in this on this plane of existence. The religious theory of "hell" is the period and time when the macrocosm frees itself from the ellipse (the hereafter) then the microcosm(s) which have not resolved their ellipse before that will be in a state of purification, and that will apparently be painful (hell).

Those that in the time period of the ellipse in macrocosm have resolved it and returned to a perfect circle (in microcosm) will be in what Western religious theory is termed heaven or paradise (in this hereafter period), while those who did not resolve their individual ellipse (in microcosm) will be in hell as the ellipse in macrocosm becomes again a perfect circle.

The Quran records those in paradise in the hereafter saying:
"This is what was given to us before."
I will have much more to say on heaven and hell in chapter 11.

Chapter 7

The Two Elliptical Lines of Descent

Historic spiritual lines of instruction can be separated into two: the Hindu, Native American, Shaman, Buddhist, Taoist line of spiritual transmission which is represented by the 1, 8 ellipse (*Figure 18*). And the "People of the Book" or Western Middle Eastern Semitic line of spiritual lore represented in the ancient Egyptian, Iranian, Jewish, Christian and Islamic line, figured in the 2, 7 ellipse (*Figure 19*). The father or source of these two racial-religious expressions is the Sumerian race or culture.

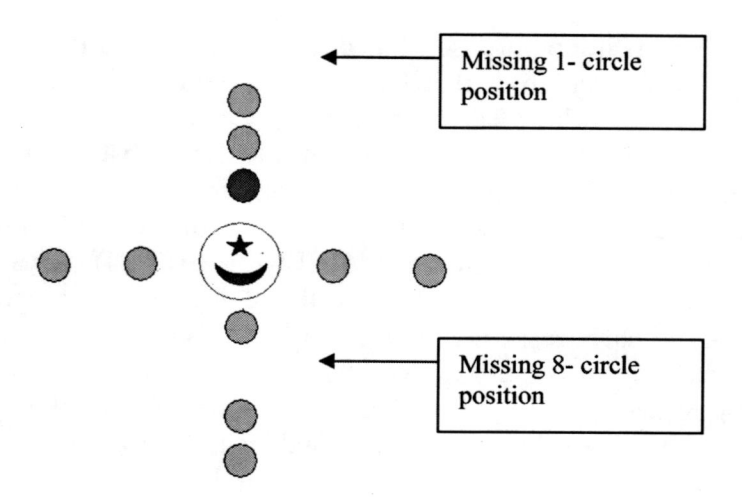

Figure 18: Representation of ellipse 1, 8 showing missing/ejected circles 1, and 8.

The Ellipse: The Fall and Rise of the Human Soul

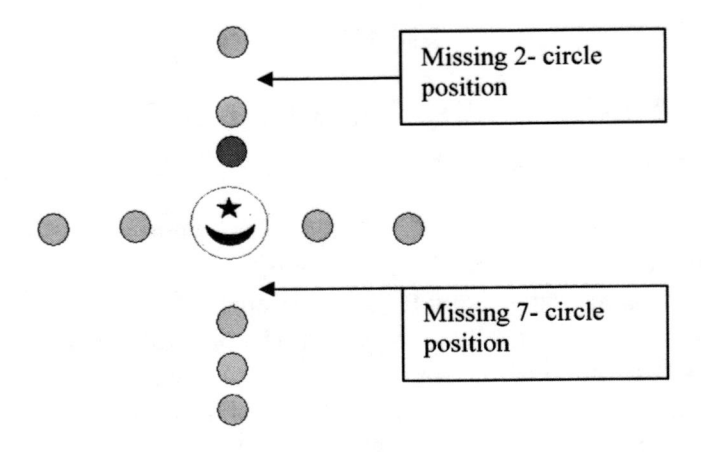

Figure 19: Representation of ellipse 2, 7 showing missing/ejected circles 2, and 7.

For clarification purposes, this double elliptical reality does not mean in the macrocosm that the ellipse is separated in two. The macrocosmic world - despite the two lines of elliptical creations - is one world or a unity that represents our holistic selves as imperfect circles, converging in history and representing two types of beings coming together to form a whole. In fact, the reality is that IN THE MACROCOSM THERE ARE (4) MISSING CIRCLES! As distinguished from the 2 separate microcosmic ellipses that share respectively two separate sets of missing circles.

History records that the two lines of descent of the fallen man clearly reflect the two different ellipse structures. That being the earthly more open Shaman, Hindu, Buddhist, Taoist man, the 1, 8 structure; in comparison to the more rigid 2, 7 Egypt, Judeo-Christian-Islamic being. Indeed, it is the two missing elements in the ellipse structures that determine the historical behavior and religious expression of the respective fallen beings, the 1, 8, or the 2, 7. I hasten to say that any objective examination of history bears this out.

The 1, 8, for example: looking at the two missing or out of line circles and what they represent to our inner natures demonstrates in this particular ellipse why they act the way they do. The 1-circle (the circle of truth in potential) missing from the 1, 8 elliptical being reflects their religious nature or expression, as the 8 or earth circle also illustrates that the 1, 8 being, in missing his inner 8 or earth circle, is reflected in the nature of the Shaman-Hindu-Buddhist religious expression. Certainly most can agree that as a whole or generally speaking this Shaman-Hindu-Buddhist 1, 8 man is open to other beliefs and more earth-centered (circle 8) in his nature. REMEMBER, A LACKING CIRCLE MEANS THAT THE PERSON IN ACTING A CERTAIN WAY IS ONLY REFLECTING WHAT IS LACKING IN THEIR SOUL. In this example, the 1, 8 circles. In other words the missing circle natures in our 1, 8 fallen being will determine how he behaves or what they lack in not having access to their 1, 8 circles.

Similarly the 2, 7 ellipse, is a more rigid person interested in enforcing monotheism, dogma, and earthly control because of the very lack of the access this being has to his 2, 7 circles; he is compensating. That's why the Western Semitic religious person, that being primarily the "People of the Book", the Jewish, Christian, and Islamic dispensation are more concerned with religious dogma and imposition of their faith on others, reflecting their missing circle natures, that being the 2, (physicalization of truth circle) and circle 7 (the Bey or Lord of the earth circle). That is why the 2, 7 "People of the Book" history reflects a concern with dogma, land acquisition, religious dispute and terminology. In both cases the missing circles reflect the activity of the given ellipse structure.

This interpretation of our being, though, should not be taken as a 100 percent correspondence in natures to this principle, so I am not at all saying that all 2, 7 people are rigid and dogmatic and all 1, 8 are more tolerant and open. There is certainly a convergence of natures and true, the 1, 8 and 2, 7 natures do seem to have similarities and a more in-depth study of the two sets of out of line circles will lend a more precise understanding of the similarities and differences in the two ellipse structures.

The historical obsession of the Abrahamic Semitic dispensation, Judaism, Christianity, and Islam, is a clear proof that they are primarily beings of the 2, 7-lineage ellipse.

Looking closely at the two divine essence circles that are missing the 2, and the 7 we see the etiology of this.

The 2-circle is the circle that holds within its essence the divine archetype of the physicalization of truth. Now what does that mean?

This circle or essence is the energy that controls this inert attribute that rules the subtle laws that bring reality to being, and in a form. This essence is missing in the 2, 7 ellipse, so therefore in their religion they have an extreme obsession with religious form reflecting what they lack in the 2-circle. What must be said here now is that all Western religionist are not 2, 7, as all Eastern religionist are not, 1, 8. However, the pronounced obsession with this by Western religious form is proof that this is their essential circle intelligence. The 7-circle only complements and reinforces this, considerably. Indeed the Bey (Lord) of the earth 7-circle is all about the inert subtle energy that rules a body of earth or spirit, consequently the obsession of the "People of the Book" - Western religion - with holy lands, tribes, favorites of God and divine real estate!

On the other hand the 1, 8 primarily Buddhist, Vedic, Shaman, Taoist being has not nearly such an obsession. However, their earthly, less form-intensive philosophy indicates clearly the lack of the 1 and 8-circle in this ellipse being. The 1-circle is the divine essence that rules the subtle reality of (truth in potential) and the 8 divine essence - the earth circle and all that that energy entails - is reflected in the Eastern religions. All one has to do is look at Taoism, Buddhism, Hinduism, or Shamanism, and contemplate these two missing circles and we clearly see that both of these types are compensating for what they lack in their souls.

Lets look at the 2, 7, first:

Remember these obsessive traits are illustrated to underlie the concerns of the particular ellipse. This is not an attempt to cast aspersions on any people or group. WE ARE RULED BY ENERGY, AND ITS CONFIGURATION - IN OUR SOULS, THAT IS EXISTENTIAL REALITY, THAT CANNOT BE GOTTEN AROUND!

Judaism: Major concerns or obsessions: God has chosen people. God's land of Israel. Religion only exclusive to the "Jews." Excluding all others.

Holy Land obsession: Jerusalem

Christianity: Major concerns or obsessions: Jesus, Jesus, Jesus! Only Jesus is the way.

They doggedly hold on to this, and have an extreme exclusivity obsession, similar to the Jews, but reflected outwardly and aggressively towards others, whereas the Jewish people reflect this inwardly. Obsession as the Jews have, with them alone as God's favorites, or people.

Holy land obsession: Jerusalem.

Islam: Major concerns or obsessions: Ardently announcing how the others, Jews, and Christians, have gone astray, while at the same time becoming (although not as much) as exclusive as the other groups. God is one, has no friends or partners, and Muhammad is the last messenger.

Holy Land obsession: Mecca, and Jerusalem!

Again, I repeat it is not my intention to cast a negative caricature of these great faiths, for they have many positive concerns and aspects of their theology, but this is an attempt to illustrate that these emphases are clearly similar, different shades of the same color and indicate the lacking essences. They are unconsciously carrying these traits unknown to them, why they act the way they do; now they know! These are indeed powerful essences.

BUDDHISM, HINDUISM, JAINISM, TAOISM, SHAMANISM

Anyone who knows anything about these religions can clearly see the earthly aspect of them, as well as the far less compulsive, linear, doctrinaire and Holy Land obsessions than the 2,7 "People of the Book."

There is certainly formalism and doctrine in these faiths, but the aspect of these parts of their faiths are not remotely as obsessive or pathological or as negative, as the 2, 7 man. There is not much history of these faiths murdering, oppressing others for their beliefs as the Western group. That is because this group is not missing the 2, 7 circles thereby they have a somewhat healthy concern with these aspects of their faith and their natures. In other words, you will not see Buddhists or Hindus going around chopping people's heads off because they do not believe in their religions articles of faith. That of course is not to say that Hindus and Buddhists have never done this before, but compared to the 2, 7 group, it's not even close. In addition, what must be taken into consideration is that there are 2, 7 Hindus and Buddhist, but not that many. Similarly there are 1, 8 Christians, Jews, and Muslims, but the 2, 7 predominate.

Also, if we study the doctrines of these faiths, they are much more open, accepting of others' ideas, not as closed, or even offended, or obsessed with getting others to believe what they believe.

On the other hand one might say that these faiths may be too open, not in a negative way, but existentially this is a fact, even though this lack of their 8 and 1-circle has a less negative impact on others as the 2, 7 man's missing circles. However, just as the 2, 7 person, the 1, 8 is compensating for his lacking 1 and 8-circle that renders the personality of their faith extremely open and earth-centered because they lack these elements, they are compensating as well.

It appears that in history the 1, 8 man was first to develop his spiritual language of return to perfection as I think history testifies chronologically to this, even though there is certainly a historical and archeological dispute of which people - the Egyptians, Sumerians or Hindus - were the first to become "civilized." In the certain history we can probably agree on, it is clear that the 1, 8 line perfected its expression in the appearance of Gautama Siddhartha or Buddha; and the 2, 7 line of spiritual return completed its dispensation on the advent of the Prophet Muhammad in 570 AD, 1200 years or so after the Buddha, as Muslims believe in declaring Muhammad as the culmination and completion of the Abrahamic dispensation.

MYSTICAL VIEW OF THE CIRCLES (ESSENCES) AND SPIRIT

Mystically, many of the major esoteric paths, Yoga, Sufi, Buddhist, and Kabbalah, represent major aspects of our circle, and the spirit. The 10 Sefirot trees of Kabbalah are the Holy Spirit. Buddhism represents the center 5-circle (in the 1, 8 ellipse) because his fundamental teachings emphasized unity (core principle) and nirvana (pure energy-emptiness) which is what the center circle essentially means. Similarly, the Prophet Muhammad reflects the light of the center 5-circle in the 2, 7 ellipse human, as the essence of the Abrahamic dispensation.

In the metaphysics of Sufism, they concentrate on activating the subtle organs of perception as a way to enlightenment. In this, they are communicating with the circle that contains these subtleties and the *nafs* (human spirit-self), the 6-circle. Similarly, the yoga philosophy represents the 15-circle in that it contains the 7-chakra wheels.

It must be noted though that in no way is this a literal denouement of the circles and spirit with these paths, since all the paths emphasize the proper balance of all the divine energies in their schools of thought. These mystical, or at minimum symbolic, associations, are just that, symbolic, and not to be taken too literally. That being said, on the other hand these associations of the major paths with these divine energies have a validity in terms of what the paths emphasize as a way to the truth.

Chapter 8

History of Religion After the Fall

A brief history of religion and its sources

Religion and mysticism become necessary because of the fall. Religion began after the fall of man, through the Evolutionary Intelligence:

Quran 2: 38 "We said: go forth from this state all of you there will come from me guidance from me whoever follows my guidance, no fear shall come upon them, nor will they grieve"

As mentioned earlier, this fall took centuries and was a long process that began in the period of time when the descendants of the transcendent Adam - a far lineage within the lower Adam - was born without the total perfection of the inner elements of the heart-soul that would have protected the fallen Adam from the machinations of the devil. This is recorded in the excellent book *The Book Of Certainty* by the Sufi Siraj ad-Din[*]

This Sufi mystical cosmological treatise describes the three levels of knowledge as, the lore of certainty; the eye of certainty; and the truth of certainty - the highest degree of knowledge.

THREE DEGREES

The Lore of certainty: This is the oral and written tradition of the exoteric law primarily, and some of the esoteric lore. This is the beginning degree of faith that all humans have access to, and

---

[*] The Book of Certainty, Abu Bakr Siraj ad-Din

can according to their own altitude, benefit from.

The Eye of certainty: This is a level above the lore of certainty that begins the opening up of the inner faculties of intuition and perception. This begins real experience of the extraordinary insights, and experiences that can carry the student to the next level.

The Truth of certainty: In the book, the Truth of certainty is described beautifully and poetically by Siraj ad-Den: the experience of Moses at the burning bush he states is an example of arriving at the truth of certainty, where all doubt has been erased, and the human experiences first hand the divine:

"And when he reached it, he was called by name: O Moses! Verily I am thy Lord. So take off thy sandals. Verily thou art in the holy Valley of Tuwa," Quran 20, 12. *

Moses reached the burning Bush and his taking off his sandals, that is, "by removing the very basis of his apparent existence apart from the creator in the two created worlds, Heaven, and earth" represents his extinction in the truth of certainty.

The above is from the book of certainty. It goes on to explain that to be in this final degree of knowledge is to be extinguished in the truth, where there is no longer any room for anything but God.

According to Mr. Din, the fallen Adam became over time vulnerable to the devil's machinations when he no longer was born with the level of the truth of certainty in his heart, only a form of hereditary perfection was the case. ᵀ

As mentioned above in chapter 2, this was the vulnerable lower Adam that fell to the machinations of the rebellious jinn in primordial times.

After this fall, a long period of settling down in the ellipse had to occur, for some strange sort of balanced-imbalance had to settle in on the human being in order for the spiritual hierarchy to begin the process of initiating the Lore of certainty - level of guidance

---

* The Book of Certainty, p. 2
ᵀ The Book of Certainly, p. 29

promised to humans. Religion had no reality in the era of the transcendent Adam before the fall of the lower Adam. Only the high-level magical practices of the jinn were anything resembling religion. However, this was for their high-energy-based civilization that needed this to balance the solar heat that their created nature demanded. There was no God or salvation consciousness in the primordial era, as we know today, for no one needed saving.

Salvation became necessary for only the rebellious jinn, and the fallen Adam. Salvation from what? Salvation when the ellipse is dissolved. Those microcosms that have not resolved their inner ellipse will be in the extremities of discomfort (hell) in the macrocosmic era of the dissolution of the ellipse - apocalypse. Much more on this is in chapter 13, "Paradise and Hell".

These rebellious jinn essentially became the reality of the "devil" on earth. Their lineage was sought and wiped out whenever possible by the hierarchy, but that was not completely done. Their heritage is still on and in the earth as well as their affecting humans through the mixing of our DNA with theirs in primordial times, as recorded in *The Book of Enoch*. Consequently their salvation also became necessary - that is, only the rebellious jinn - not those who remained loyal to the hierarchy; they evolved to the transcendent Adam stage if they remained loyal, so did those who had no problem with the judgment period.

The first guidance that came to humans was through one known by different names, according to different races. This was Enoch to the Hebrews, Idries to the Arabs, Hermes to the Greeks, Thoth to the Egyptians, and Mercury to the Romans.

In addition, he is known as Hermes Trismegistus, the thrice great, responsible for the religion that has come down to us known as the Hermetic tradition. Hermes was their first emissary, and a powerful and great one.

HERMES

Undoubtedly, this ancient figure was a composite of a series of teachers who spoke the same spiritual language, and who came down to various cultures.

The Hermetic tradition was a non-sectarian mysticism that allowed all cultures to adapt this lore for themselves, and create traditions suitable to them.

Obviously much mythology was built around our great spiritual ancestor, and many great mystical treatises are ascribed to Hermes. The ancient document, *The Emerald Tablet*, seems to be the basis of the Hermetic philosophy. This mysterious document also contains the famous: As Above, So Below aphorism.

The famous *Poemandres* is also a document ascribed to Hermes.

Science, magic, religion, philosophy ... this eclecticism was no doubt designed to appeal to a wide variety of types from the jinn to the simple human who needed moral guidance. Primarily, though, it is clear that Hermes was the teacher of teachers, and his material was high level mystical knowledge that was the beginning wisdom bequeathed to humans after the fall.

This wisdom became the catalyst for future religion, and mystical paths, after the fall. We could call Hermes the father of religion, mysticism, science, and magic.

He was an itinerant traveler; never ceasing to spread wisdom to whichever people he appeared amongst, according to their level of understanding. He was the super adept, Hierophant, who was the first organizer of the lore and knowledge given to humans for their return to perfection. This included all kinds of knowledge that span the spectrum of metaphysics from ceremonial magic to moral guidance.

The Quran refers to Idries in this way: "we raised him to a great height." Only Jesus is referred similarly to in the Quran. This is very important because, Jesus being the chief archetype of the classic teaching master, Idries is the chief archetype of the teacher

Hierophant, a level above the teacher archetype that is in a sense the teacher of the teachers.

Hermetic lore was the foundation of the knowledge that would be the fruit and nutrient for all future teaching masters in the field of mysticism and religion on this earth.

NOAH: THE AGE OF SEPARATION

This era was an attempt by the hierarchy to separate the wheat from the chaff, so to speak. The elements in humans they deemed too destructive (because of the fall) to be suitable for guidance. The biblical reference to the deluge of Noah is a symbolic reference to this era. This is not a literal true story, but an allegory, as is every tradition of a deluge-like phenomenon, being a reference to this period after the fall: in reality, making the distinction between the guilty and those deemed partially fit for guidance.

ABRAHAM: THE GATHERING

The first indication of a codified 2, 7-ellipse intelligent religion in the West was to be the shepherd Abraham and his family, who according to the Bible was situated in what is today Eastern Turkey, in the land of the Chaldees.

At about the same time of Abraham, 1500 B.C., the 2, 7 manifestation of the Arian was manifesting in the religion of the Iranian Prophet, Zoroaster, who some believe was the origin of Semitic monotheism.

The environment according to the Quran and Torah was that Abraham was in a religion of Idolatry that the populace practiced at that time. It is recorded that Abraham and his progeny Ishmael and Isaac were dissatisfied with this pagan religion and sought something else.

Some of this Idolatry was an archetypal offshoot of the practices of the jinn before the fall, and before the interaction of the jinn races with the lower Adam. These practices were most likely based on the high magical practices of the Atlanteans and Lemurians. This legacy was inevitable, as well as the misinterpretation of the fallen man of this by translating his inward inclination towards magic to a pagan religion. This was a heritage of the confusion of the interpretation of the archetype in the fallen man because of the mixing of the jinn DNA with that of the lower Adam.

Abraham was a descendant of the part of the lower Adamic race that did not fall. He was part of the fallen human that was closest to the higher Adamic race, therefore was able to grasp the new science of religion inspired in him by the now hidden higher Adamic adepts that were the guides of the fallen race.

Mystically it is only that archetype in the fallen human, the lower Adamic archetype that did not fall, that allows us to understand this new religion - not from the standpoint of religion, but from the perspective of what the religion is aiming at: the return of the lost soul to its past perfection.

The religion itself was a foreign element not even encrypted in our archetypes, because it did not have any existential existence before the fall of humans, so how could there be any archetype? This is the reason that people have made religion more a magical thinking apparatus than a moral pursuit. The magical thinking is from the jinn archetype, since any real magic was only practiced by them in primordial times.

The high level exercises of meditation, concentration, and contemplation are from the higher Adamic archetypes.

Therefore, the early religion became a pursuit of salvation of the race of Abraham in the sense that only the aspect of the lower Adamic race that did not fall became the object of the religion of Abraham. In other words, Abraham was commanded to seek out only that being who was exhibiting his inner archetype of the lower Adam who did not fall. This is the reason that the religion initially was restricted; this is also the reason why the early religions in Hinduism, and others, had that element of hierarchy and insular sectarianism, because only certain people who were inheritors of the archetype in question could benefit from this new religion on earth.

Indeed, it is recorded that all the early religions had an enormous amount of that kind of insular thinking that translated into hierarchical sectarianism. This initially was part of the science of return, because the fallen Adam who mixed directly with the jinn races was not psychologically fit to understand the religion, so he was avoided, and practiced the early paganism of the people of Abraham - in the Chaldean part of his history.

His later sojourn in the Middle East was to seek out this archetype in the people there. They were in tune with Abraham inwardly. This culture eventually would become known as the "People of the Book" - the heritage of Hebrew, Christians and Muslims who are the heirs to the archetype of the lower Adam that did not fall.

The entire recorded early history of religion in the Quran and Torah is about the separation of the archetype of this lower Adam that did not fall, and the lower Adam that did. The people who perished in the Noah story, those who perished in the Lot story, and numerous examples of the believers and disbelievers tussling for God's favor, and disfavor, IS AN ARCHETYPE WAR! The spiritual chronicles allude to this theme often, describing these groups as intrinsically enemies of each other, as recorded in the stories of the prophets saving some, while others meet destruction. This is a constant theme in the holy books.

This is the reason that, other than Buddha and Muhammad, most early religions had this element of hierarchical sectarianism that only ended formally in the dispensations of these two, at least on an esoteric level. Even though the habit of the exoteric/esoteric aspect of religion is still predominant in the world, a heritage of this early period of separation of people according to inner archetype still pervades today's cultures, though sublimated in other conflicts.

ISHMAEL AND ISAAC: THE SEMITIC PATRIARCHS

Abraham became the seed of the separation, and his two sons, Isaac and Ishmael, became the sources of the spiritual information given to this lineage. Through this Semitic seed was produced Muhammad, from the seed of Ishmael, and Jesus from the seed of Isaac.

The Hebrew prophets of the Isaac lineage are better known than the Ishmaeli Prophets, but the Quran records many of the Ishmaeli Prophets, and as with Hebrew history, it was a separation history, where the believers were separated from the disbelievers. This was another result of the 2, 7-ellipse lineage of the majority of the "People of the Book", as well as the archetype war.

The similarities of the Isaac branch and the Ishmail group is striking. One would be hard pressed to tell the difference from an orthodox Jew with a yarmulke, and the devout Muslim with the takir on his head.

It is a tragedy of immense proportions that these people have through the years found it - as in today's Mideast conflict - necessary to fight and kill each other for land, dynastic reasons, and for some amongst them, doctrinal religion as an excuse to kill their brothers.

The religions that sprang from Abraham: Judaism, Christianity and Islam are as close to each other doctrinally, and philosophically, as any in the world. That fact does not prevent many adherents to these religions from indulging in divisive and destructive sectarianism, despite the heroic attempts at inter-faith unity by some.

The New Age movement, however maligned by some, should be given immense credit in helping to bridge the gap in this regard.

For the people of Isaac, the people of Ishmael, that is Jew, Christian and Muslim, are without a doubt from the same source, and have always, and will always share the same God, so by that God's love, should know that they are brothers and sisters.

From the crusades, the brutal assault of Christendom on the

Muslim lands, to the slaughter of the Cathars - a Gnostic Christian sect of the 12 the century - by the order of the Pope, to the Muslim slave trade of Africans, and the cultural aggression of the Arabs on them. Now we have the destruction and oppression of Palestine by the "chosen people of God." These deceitful activities by the so-called "People of the Book" on innocent helpless people is an abomination. There are numerous other incidences of this barbarity by these so-called religions, of today, and the past, but it would take an encyclopedia to list them. Even today these same groups are separated and can find no way to unite behind what they have in common. They still maintain their sectarian ideas of superiority of themselves, and their creed over others, and can find no way to extricate themselves from this deadly poisonous mentality, that they have the audacity to blame on God, and their religions!

These people are a disgrace and abomination to God, man and all that ascribe to any real decency, goodness and unity in the human race. Sectarianism should be avoided by anyone who wishes to advance spiritually, and the lame reasons that this group or that group has been around for centuries is no excuse to follow his or her poisonous ideas of division, superiority, and sectarianism.

He who separates the human race is separate from God!

THE TWO PATHS

The spiritual systems in the world seem to have a strange dichotomy. Millions of Westerners in traditional Islam, Christianity, Judaism, and the vast majority of the sects spawned from these faiths, have only a vague belief in a heaven and hell concept after death.

It has always been interesting to me how knowledge of the inner path has evolved in the East as in comparison to the West.

Hinduism (yoga) and Taoism in its best light seems to put the attainment of enlightenment at the forefront of its system as certainly, I think, at least all the systems of Buddhism also put the attainment of enlightenment at their core.

Whereas the Western mystical tradition seems to have split their systems in two:

Islam/Sufism
Judaism/Kabbalah
Christianity/Gnosticism/Mystic Christianity

The division is an interesting phenomenon, for example:

Kabalists claim to be Jews generally, but are not accepted or hardly known by the average Jew, Madonna notwithstanding!

Similarly with Sufism that claims a connection with Islam but hardly any conventional Muslim knows that Sufism exits, let alone what it is all about.

In Gnosticism it is the same situation, as well as for other forms of Christian mysticism.

In addition, the exoteric systems of the West seem to have a hostility towards the esoteric systems attached to their faiths. Whereas I do not think that such hostility exists to any degree in the Eastern and Far Eastern traditions. Why the split?

Again the 2, 7 1, 8 difference is probably the reason for this, although there are other possibilities that may add to this dynamic.

The overall phenomenon of elitism, that although shunned in theory in these New Age, non-hierarchical times, is dominant in all mystical systems, but primarily hidden, even though this is downplayed for obvious reasons. It just does not look good. But it is a fact of life that there is a caste system that separates us. But the contention here is that it is based on the inner association or leaning towards the inner archetypes that determine this hierarchy. There is not an iota of religious or racial superiority behind this. Those closest to the inner types nearest the transcendent Adamic strain are the elite because they are closer to the perfect nature and that and that alone allows reality to give them preference or favor. This is a hidden phenomenon primarily, but does show up clearly in the historical fact that most of our greatest mystics and religious teachers were middle class to well off materially. God seems to take care of his own, despite the condition of the world. Many do not like hearing this, but just research it, and it will be confirmed.

This is not a bad thing, in itself, just reality, for even Buddha said, "I will return to teach for those with a little dust in their eyes." He didn't say he would return to teach all!

THE WESTERN RELIGIOUS EPOCH TIMELINE AFTER THE FALL

Ellipse 2, 7 dispensation: Began with Adam entering the age of the creation of the ellipse. This period in time is uncharted, the primordial era, in which began the fall that initiated historical time for Adam.

Idries/Hermes/Enoch: The age of initial learning and dissemination of seeds of guidance.

Noah - the age of separation and the beginning of the science of return: This antediluvian period, and after the great flood, is all about the beginning of the mystical and religious sciences.

The story of Noah: "Noah was the first tiller of the soil. He planted a vineyard; and he drank of the wine, and became drunk, and lay uncovered in his tent." Noah's son Ham saw his father naked and informed his brothers, who covered Noah while averting their eyes. Noah awoke and cursed Ham's son Canaan with eternal slavery, while giving his blessing to Shem and Japheth: "Blessed by the Lord my God be Shem; and let Canaan be his slave. God enlarge Japheth, and let him dwell in the tents of Shem; and let Canaan be his slave."*

Noah died 350 years after the Flood, at the age of 950, the last

---

* Genesis 9-20

of the immensely long-lived antediluvian patriarchs. The maximum human lifespan, as depicted by the Bible, diminishes rapidly thereafter, from as much as 900 years to the 120 years of Moses within just a few generations.

This is allegory representing Noah as the first intelligence to attempt to re-develop the science of the essence or Garden of Eden states that had been lost. The reference to Noah being a tiller of the land and planting a vineyard is all allegory for his inward traveling or his developing spiritual states and stations.

Egyptian and Sumerian religion: The beginning of the organized dynastic creeds that were the progenitors of Western religion.

Abraham - the establishment of the covenant:
This began around 1900-1700 BC and includes his two sons, Ishmael, and Isaac. This is the period when the seed of the religious - exoteric and esoteric science - was to be established by these patriarchs.

Isaac and Ishmael: Began the two lineages of the "People of the Book," the Semitic Jews, in the case of Isaac being his descendant, and the Semitic Arabs in the case of Ishmael.

Moses - the age of the law: Estimated to be about 1400 BC. The establishment of the law of the exoteric and esoteric.

Much of the shenanigans of the Hebrews reported in the Bible are really an allegorical reference to them separating the law into the exoteric and esoteric. This dynamic dichotomy is based on the 2, 7 ellipse. This ellipse in history, unlike the 1, 8 ellipse, has had a hard time because of its fallen nature reconciling the two seas of the exoteric and esoteric. In the West, it is because of the 2, 7 syndrome of this ellipse therefore separated the exoteric from the esoteric in many Western branches or sects of the law. Gnostics,

Kabalists, mystic Christians, Sufis kept the esoteric and the exoteric in a unity. They were more than likely the 1,8 people amongst the 2 and 7 branch of ellipse. Recall, in the West the 2, 7 ellipse predominates.

Jesus - The age of Return 0 - 300 AC: The first large group of humans who by reality were of the transcendent Adamic nature, and the small branch of fallen Adams who did not fall, restored themselves to the perfect circle. This rich, challenging, but mysterious period recorded many humans returning to their lost nature in this world. This occurred for about 3 centuries after Christ but slowed down and degenerated at the advent of Catholicism and the Roman Empire perverting and upending the true teachings of the Christ archetype.

Muhammad - The age of completion 700 - 1500 AC: The age of the ultimate perfection and return of many humans to the perfect circle and even beyond to the ultimate perfection of the human species in that many in the microcosm have evolved to completion and the advent of the Sufis is an indication of this. This rich era produced many people who resolved their ellipse, ending in 1500 at the advent of the Gog and Magog period and Dajjal (Antichrist), initiated symbolically by the advent of Christopher Columbus to the "New World" which precipitated the holocaust of the Native Americans, the enslavement and destruction of millions of black Africans, and the colonization of many other cultures by predominantly European nations.

1500 to present: The Dark Age (Kali Yuga) in Hindu cosmology: Kali Yuga represents the age, the last of a cycle of four, in the Hindu eschatology, when the human civilization degenerates. Equally in Islamic, and Christian prophetic traditions this era represents spiritual darkness; crass materialism, and the age when the Antichrist energy will be widespread.

The negative energies that the ellipse produces are magnified and will eventually overturn and dominate the world in this modern era. The delusion that modern science and technology is "progress" can spell disaster for the world if material and scientific advancements are not balanced by moral and spiritual development.

Later came the period of modernism, and postmodernism, two cultural/philosophical movements that began in the late nineteenth and early twentieth century in the case of modernism - and in the late forties, after World War II, in the case of postmodernism. In this period, primarily Western thinkers began to question all religious and authoritative institutions, which included metaphysics, as well as authorities in academics, politics, and art. Postmodernism is on one level an extension of modernism, at the same time a reaction against some aspects of modernism. These cultural patterns spread widely in Western civilization.

EASTERN RELIGION

1, 8 dispensation

The more earthly centered philosophy can be seen in the Eastern religions with their emphasis on Avatars in Hinduism as their "prophets" of God or Gods, and the Buddhas in Buddhism.

This history is very different from the more linear and well-defined Western religious history.

This history includes:

Hindu/Yoga 1000 BC

Jainism 600 BC

Confucianism 500 BC

Buddhism 500 BC

Taoism 300-400 BC

Shinto 300 BC?

These are all legitimate expressions of the guidance coming to humans to return to the perfect circle and beyond. All of these religions, though, do not necessarily spring from each other, in fact, they are all derived from individuals who may have been adherents to the religions of their day because of birth, but who nevertheless brought a new vision separate from those religions that existed before them. For example, it is clear that Buddhism is not derived at all from Hinduism, even though Buddha grew up in a Hindu culture. Certainly, Buddhism has similarities to Hinduism but a close examination will demonstrate they have major differences that indicate a very minor influence of Hinduism on Buddha.

Also the beautiful philosophy of Jainism, the religion that sprung from India in the 6th century BC is not derived from Hinduism either, although like Buddhism has similarities to Hinduism. Certainly the Buddhist tradition has similarities to Jainism, as they do with Hinduism, as they all have striking differences also. But one can see the relationship between these creeds primarily due to the fact of their geographical affinity.

Each of these creeds has a close relationship with the Shramana tradition in India of the wandering monk seeking truth through ascetic practices.

The world has a deep debt to the very rich, fine, and beautiful religious cultural environment of India. This mystical and religious environment that produced these creeds is without a doubt equal to the Jerusalem-Mecca syndrome of holy places in the Mideast; it is just that the great country of India includes in it their own Mecca and Jerusalem. That lends our world of extreme problems a happy face on the traditional unhappy one we mostly have in this challenging world of suffering.

So indeed our world owes a great debt to the Indian people for enriching us in this regard, as we also owe a great debt to the equally spiritual people of the Mideast who brought us the great religions that have, despite the problems they have within them, enriched us all.

In its original form there is hardly a more sublime philosophy than Taoism ever produced by humans. There is no doubt that this creed is from the highest heaven, and is a gift to the fallen man, that if he abided by it would return him to his highest self.

This is a 1, 8 expression of the ellipse certainly, as most Eastern and Far East religious phenomena are. There is no doubt that this is the yin of the yang of religion; that is, most of the Eastern and Far Eastern religions, as opposed to Mideast and Western religions.

It is clear in studying the history of the religious dispensation of the East and Far East that the spiritual hierarchy wisely guided these people through their own inclination of their soul condition. All this is a great lesson in human and divine nature for us.

Most of these religions are more homogeneous than their Western brothers, in that most of them have, along with their cosmologies, a philosophy of enlightenment that is alien to the Western exoteric tradition. One would have to go to the esoteric traditions of the West to find similar philosophies of enlightenment.

There is no stark esoteric, exoteric separation in most of these Eastern creeds as there is in the Western creeds.

Most of these systems have a healthy emphasis on the attainment of enlightenment, or liberation from the world.

Unfortunately, though they, like their Western brothers, have to a great degree degenerated into magical, sky-God exoteric superstitious beliefs in many aspects of their faiths.

"WITH A LITTLE DUST IN THEIR EYES"

As mentioned above in the Western religious section, the concept of an "elite" must be dealt with honestly in here as well, and that is the obvious elitism inherent in all religion and mysticism. Again, this elitism is not based on birth, race, or religion, but the hidden inner inclination of certain humans to the

History of Religion After the Fall

archetypes that rule and influence our religious aspirations, and ability for success in the endeavor.

Of course, we all know about the crass, Hindu caste system that all rightly condemn, even most astute Hindus themselves. Similar to the way profane people abuse and distort truth, the idea that comes through primitive Hinduism is a distortion of the true reality of the elite of God. Fortunately, this system seems to be slowly dying out in India. Nevertheless a true caste system exists in the Eastern mystical paths, as they do in the West, albeit hidden and almost impossible to decipher.

The fact is these people themselves (the *real* elite) have not an iota of elitist chauvinism in their heart. Any claim of being part of some elite by ignoramuses like the abominable Nazis, and other racial elitist, and religious sectarian abominations that come from the lowest common denominator in the human species, has no resemblance to this concept, that solely is based on a hidden reality. This is rarely seen but by the most advanced seers into the human heart to any degree. And even here, it is a rare person who is favored with this knowledge of who the elite really are.

Certainly, Buddha's statement, made after deciding that he would teach after he attained enlightenment: "I will return for those with a little dust in their eyes," is a statement to what I am referring to.

This is again, not equal to the profane worldly doctrines of elitism that have distorted this simple truth. For any favor these elite get is only based on the mystical system of dharma, in that the teachers and the unseen grace of cosmic science detect that these will have a clear way to success on the path.

Nevertheless, heroes like Buddha offered his system to all - and certainly despite the truth of those drawn nigh to God (the elite) - the lowest of the low and all humans in any true legitimate mystical system are invited to advance to success according to their aptitude. Indeed, no one knows this better than this writer who has come from the lowest depths of the human condition to the highest truths of the universe by some mysterious grace.

THE SHAMAN AND THE NATIVES

From the practices and beliefs of the Native Americans from the North and South American continents, the shaman witch doctors of Africa, to the Aborigines of Australia, we find primarily a 1, 8 dispensation of religious and cultural beliefs that are as ancient as any of the dynastic creeds out of the Indus valley, or Mideast cultures of Egypt and the Sumerian. Indeed, these people were more than likely the pioneers that began return to spiritual knowledge and reunion with the inner soul. In addition, they are an embodiment of the culture of the primordial times, much more so than Western and Eastern man is.

Many of the practitioners of these religions did not really consider themselves as practitioners of religion, but only practicing a natural spiritual form of living. This indicates a closeness to the primordial times when the Adamic races were in paradise on earth, before the fall.

The religions of what some consider native peoples is, despite what people think, often times are far more advanced than Western or Eastern religions. This again indicates a closeness to the true Adamic period in of our history. Though, like other religious phenomena they have to some degree perverted their guidance.

The Orisha concept of the Yoruba people of Africa is very similar to the essences of the Garden of Eden, as many of the indigenous religious concepts of the Mayan epiphany are superior or equal to Western concepts of science and religion.

Chapter 9

The Ellipse and the Myth of Jesus

Our historical Jesus story is entirely mythological and allegorical. In fact, what religionists do not understand is that its essential value is mythological, and initiatory. It has no value or reality as a doctrinal phenomenon, or as a literal history. In fact, to emphasize its doctrinal aspect has led, and will always lead, to division, ignorance, and intense suffering.

The ellipse or corrupted circle, our metaphor for the fallen microcosm and macrocosm, is also the chief symbol and reference in the allegorical tale of Jesus. The very essence of the story - that is, the crucifixion - is a reference to the phenomenon of the ellipse; in the myth of Jesus as sacrifice and symbol of the fallen lower Adam being sacrificed (that is practically all of humankind) as created ellipses. Indeed, even the myth about the two thieves alongside Jesus at Calvary is a reference to him representing the ellipse in macrocosm and the two thieves alongside him representing the two microcosmic ellipses in the 1, 8 and 2, 7 circles. Consequently, on one level we have in the great myth a clear illustration of the dynamic cosmic events surrounding its creation. In fact, the Christian cross is the symbol of the corrupted circle.

This is further confirmed (that is the real occult meaning of the cross) by the prophecies of Muhammad that speak about the return of Jesus, and that he (Jesus) on his return "will break crosses and kill swine."

This does not mean that Jesus will likely go around literally breaking crucifixes, and killing pigs. This reference is to the return of Jesus (the Sufis), being a counterbalance to the ellipse (crosses) and the negative energy of the ellipse (swine) can also be a reference to the fraudsters of religion and politics who exploit the ellipse for personal gain.

Breaking crosses means mystically that the Jesus archetype (second coming) that exists in this world, primarily exemplified by the Sufi Master, and enlightened Guru is one way that exists today to break free of the ellipse, and killing swine is the person controlling their lower soul in order to assist in accomplishing this freedom.

In addition, the mythological chronology further confirms our cosmology in the archetypal persons of the Mary Magdalene, as the representative of the ellipse, and Virgin Mary as the representative of the perfect circle in heaven. The two mythological figures then are not only coincidentally namesakes, but our fallen supposed reformed "prostitute" who in history as recent research is now revealing, was a mystical adept of high rank, is the very archetypal symbol of the ellipse, as Virgin Mary is the archetypal essence of the perfect circle. Remembering our earlier chronology that "hell," the ellipse - corrupted circle - is from the perfect circle - paradise: That is, the Mary Magdalene myth is not only our symbol of the ellipse, but an emanation of the Virgin Mary.

Furthermore, it is important to reveal here a secret about the historical myth that has been confused or hidden. That is that the Virgin Mary was in fact the mother of not Jesus, but WAS THE MOTHER OF JOHN THE BAPTIST whose lofty state is referred to in the Quran: "we have made none his equal." John was in fact the center of the universal perfect circle or higher Adamic man. He reflected the light of the higher central sun that is source to our solar system. He was the center light of the 5-central or perfect sun where our 4-circle ellipse is from. Consequently, he was the son of Virgin Mary, or in spiritual technical terms: HE REFLECTED OR WAS CAPTURED BY THE LIGHT OF THE HEAVENLY HOST, REPRESENTED BY THE MYTHOLOGICAL FIGURE IN THE VIRGIN MARY. Indeed that is why the New Testament has John initiating Jesus in the order of the sun, the highest mystical order in the universe. Moreover, ironically, the real mother of Jesus is not the Virgin Mary but Mary Magdalene! Jesus was in fact the lower center of the ellipse (solar system) as the mythological events illustrate. That's why he was to be symbolically sacrificed. This detail has no true physical reality, its truth is spiritual, in that, John was the first reflector of Virgin Mary's light, not really her biological son, therefore he was son of Virgin Mary spiritually, as Jesus could have very well been her biological son. When he inherited John's mantle, reflecting the light of the center of the universe, he become his mother's spiritual son, as well as her biological son.

It seems the very essence of the Christian myth is the story of the ellipse. The major elements of the myth confirm this: The ellipse is 13, Jesus had twelve disciples and he makes 13.

Virgin Mary as "Mother of God" representing the perfect circle as Jesus' "Mother" (remembering that the perfect circle is the source of the ellipse). That is Jesus as the center of the ellipse as in reality the converging mythological Mary figures (converging in a sense to both be Jesus' Mother). Even though the Virgin Mary was mythologically the "Mother" (source of light) of John the Baptist first.

Mary Magdalene in her role representing the "harlot" or ellipse, being reformed by Jesus (that is the messiah reforming the fallen world).

Jesus "dying" by crucifixion represents the suffering of not him alone, but most of humanity, whom the EI has sacrificed, and allowed to suffer for the cosmic intention of being a host for energy transformation.

His resurrection from the dead represents the fallen ellipse returning to the perfect circle (transcendent Adam) "Father in Heaven" to be raised further to completion, that is, even beyond the rank of perfection. Therefore, the myth tells us that the entire fallen Adamic race is destined like the mythological Jesus to be resurrected from the dead (ellipse) and returned to our original perfect nature and even beyond perfection to completion; that is the reason for the myth of the divinity of Jesus. It is a symbol for the return to divinity of all humans.

The aphorism "Render unto Caesar what is Caesar's and unto God what is God's" is a reference to the Gog and Magog negative energy that the ellipse produces that has to be fed on a low level, and the subtle light of God that has to be reflected and served on a higher level.

John the Baptist being beheaded at the behest of the daughter of the Jewish Princess Herodias is a allegorical reference to the perfect circle being ellipsed by God as represented by the wild fiery Goddess (Kali, Diana - forces of cosmic nature) symbolizing the dynamic that enabled the creation of the ellipse from the perfect circle.

The Judas allegory represents the original jinn of Lemuria and Atlantis, as they betrayed the lower Adam in primordial times. Similarly, Judas betrayed Jesus to the Romans (Gog and Magog - negative - energy) reflecting the ancient myth in the updated form.

We see then the myths and symbols of one of our major religions informing us clearly in the central theme in their mythological doctrine, that is the crucified Christ (as symbol of fallen Adam) on the cross suffering is in fact the sacrifice of all of us (for a time) in this very painful primarily negative experience in the great imperfect circle.

With an added lesson to us that these cosmic ARCHETYPICAL MYTHS ARE ALWAYS PLAYING THEMSELVES OUT IN SOME FORM, IN THE PANORAMA OF THE WORLD.

The reflection of heaven's myths has to be played out in our world, particularly by a prophet of God. Everything they do, however mundane or lofty, reflects heaven's acts, past, present, and future. That is the reason why Jesus and his disciples, as a pantomime, went through the myth of the crucifixion. Jesus had no choice, as a light bearer of the ellipse circle. If Jesus had gone to China trying to run away from this, it would have followed him there in some form. This is the meaning, also, of the story of the

Prophet Jonah, who tried to run away from this but as the Biblical and Quranic story illustrate, it was impossible. The sperm whale, the huge sea creature that swallowed Jonah, represented the creature of the Ocean that reflects heaven's all-embracing active power that is bound by its own light, and becomes more active as one tries to flee from it.

In the lowest world we are all reflecting what occurs in the highest world, we have no choice but to reflect its reality, as determined by the mirror of this world as it reflects heaven.

The biblical aphorism "I and the Father are one" is an indication that the tale of Jesus not only confirms our idea of the ellipse, also it appears to confirm the reality of the six natures. Jesus by example, and words, is communicating to us how the universe works (I and the Father are one; father-transcendent Adam). He in conveying his union with his higher Adamic nature (father) and is thereby expressing the occult phenomenon that he is inwardly practicing mystical science. This is explained intricately in chapter 14, Return to Perfection.

The father is the transcendent Adam that ALL humans are descended from. This celestial reality is one of the reasons of the errant doctrine of the literal "son ship of Jesus," and the idolatrous view of Jesus Christ that historical and conventional Christianity has fallen into: Mixing this cosmic mystical reality - the divine nature of the transcendent Adam, and that our source or father is that Adam, not God in a literal sense. Technically and literally speaking ALL prophets and sages are transcendent Adams, not fallen or lower Adams. Therefore, they can be called literally the sons of the unique first Father in Heaven - the Liegelord Adam.

The doctrine of return to "God" is through the cosmic nature and archetype - transcendent Adam. This is the condition we technically had before the fall as lower Adam - direct descendant of the higher Adam. We must return to that altitude before we go on to resurrection, the topic covered in a later chapter.

OTHER MYTHS

Many people have commented on the importance of the numbers 12 and 13 in religion, and mythology. The mythologist and Egyptologist, Gerald Massey, Archarya S, author of *The Christ Conspiracy*, and the movie *Zeitgeist* point out the similarities in all the ancient mythical heroes and the number twelve and thirteen. It seems all these figures had 12 disciples and the hero makes thirteen. In fact, the number twelve comes up in various forms in these religious systems:

From The Christ Conspiracy, Archarya S writes:

"In reality, it is no accident there are twelve patriarchs, 12 tribes of Israel and 12 disciples, 12 being the number of the astrological signs, as well as the 12 "houses" through which the sun passes each day and the 12 hours of day and night. Instead like the 12 Herculean tasks, the 12 "helpers" of Horus, and the 12 "generals" of Ahura-Mazda, Jesus' 12 disciples are symbolic for the zodiacal signs and do not depict any literal figures who played out a drama upon the earth circa 30 CE"[*]

This is the reason many researchers associate all these beliefs with astrology - the sun moving through the twelve zodiac signs. In there opinion all the history behind these figures, including Jesus, is all made up, or "carnalized" and the theologies are only astrology mysticism.

The thing that ties all these mythologies together is the aspect of thirteen, and the twelve disciples or planets, houses of the zodiac are in some form serving the mythical hero, who numbers the thirteenth element in these systems. Certainly one could see

---

[*] THE CHRIST CONSPIRACY: The Greatest Story Ever Sold, Acharya S, pp. 166-67

how these researchers could deduce that they are derived from astrology. However, if we look closer we see nothing but the ellipse theory of the 12 circles of the ellipse and the 13th core 5-circle.

Astrology was never a secret therefore why mythologize it? The researchers have missed the REAL secret that is the existence of the ellipse, or circle in the universe, inwardly and outwardly. The hero is the one who reflects the light from the 5-source circle on our essence-circle. It is acknowledged that the 5-core circle is indeed the sun, and the other circles are the outer planets that circle the sun (save that in the ellipse eschatology there are two circles inside of the 5-circle). Now granted astrology is about the 12 houses in the zodiac, not the planets, but certainly, these mystical systems could be referring to another layer of being that astrology is but a cover for. This is the meaning of *Secrets of the Cosmos*, the subtitle of the book.

As mentioned earlier the 5-center circle is the axis of the solar system, as well as the axis of the inner essence, in which along with the 2 inner circles, 15, and 6, are the core of the human being. All the major metaphysical teachers were at one time or other reflectors outwardly and inwardly of this center circle. The mythologist have always missed the true meaning behind this and only relate it to astrological theologies, not the secret inner mysticism revealed here. In fact, many of the masters themselves, and the participants in these religions, were not very aware of these inner and outer celestial lights that is described here in the ellipse theory, therefore certainly the well meaning researchers could not decipher the hidden reason behind the concentration on the numbers 12 and 13 in religio-mystical theologies.

The Jesus mythological eschatology captures the essence of the ellipse theory, though not even known to the practitioners of the religion themselves! For the fact is one does not have to be aware of this; the reality of the universal order will always, whether we are conscious of it or not, reflect reality as it is, and will shine its truth through any phenomenon.

Chapter 10

The Six (7) Natures

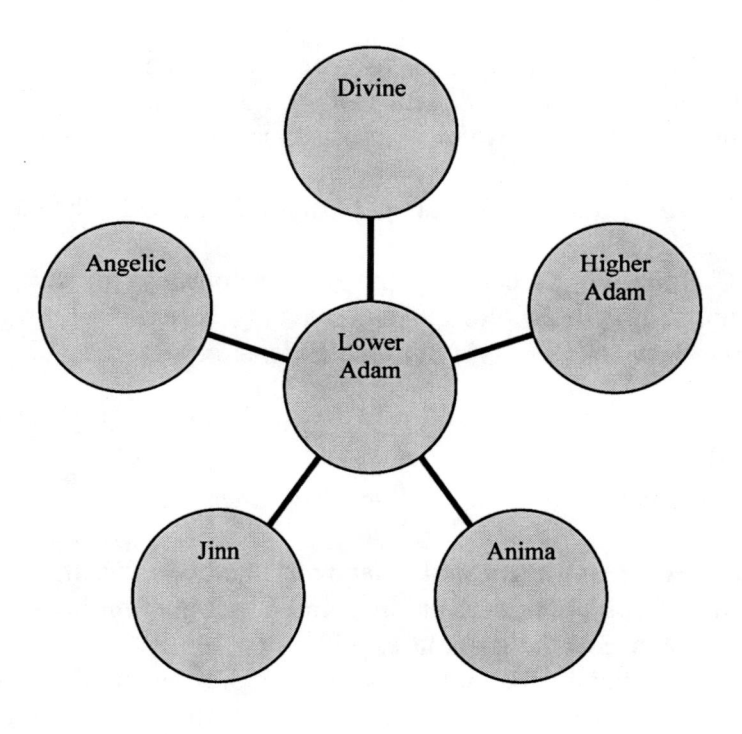

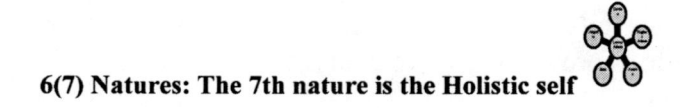

Diagram: Lower Adam (center) connected to Divine, Higher Adam, Anima, Jinn, and Angelic

6(7) Natures: The 7th nature is the Holistic self

1. Divine: The reality of the essence and spirit of all things.
Feminine: Feeling *Masculine*: Energy

2. Angelic: That which controls all reality by light and power.
Feminine: Light *Masculine*: Power

3. Higher Adam: The celestial and earthly sciences.
Feminine: Cause: *Masculine*: Effect

4. Lower Adam: The primary clay-like focus of all consciousness. Concerned with well-being, tranquility, peace and stability, it chooses when the organism reaches comfort and balance.
Feminine: Stability/Maturity *Masculine*: Growth/Change

5. Jinn: The secondary hyper-consciousness of the lower Adamic nature; that which extends reality of being.
Femenine: Stillness *Masculine*: Extension

6. Anima: The reality of the form of the unit intelligence (being).
Feminine: Nurture *Masculine*: Consume

The seventh nature is the nature of the holistic self, which at its full creation contains 6 natures, and at completion involves the perfect balance of the six natures within it.

Our mysticism that revolves around the return of the fallen circle - ellipse - to the perfect circle is what this affair of return is all about. All the symbols in the mystical cosmology of the allegorical tales in the Bible, Quran, and other mystical formulations prove this, when analyzed more deeply.

For example:

The Sufi mystic numerology called the Abjad is an ancient system of divination that is based on the Arabic alphabet. Each letter in the alphabet represents a number, for example, the word Sufi is broken down:

The Six (7) Natures

S=90
W=6
F=80
Y=10

This translates to the number 186, which in turn is broken down accordingly:

100=Q
80=F
6=U

This decodes in one arrangement to the word FUQ that means above or transcendent.

If we now take the two names Adam and Eve as they are in Arabic: ADAM.

A=1
D=4
A=1
M=40 =45

EVE (HAWWA, in Arabic). [One 'A' sound is not a letter so it is not counted.]

H=8
A=1
W=6
W=6 =21

Adam and Hawwa = 66

The point here is that the Arabic word for God is Allah, that also equals 66:

ALLAH
A=1
L=30
L=30
H=5 =66

Decoding this by mysticology, the word God in Arabic or Allah coming out to be 66 is the perfect designation of the 6 natures balanced along the male/female polarity that is one meaning of the 66. As the 66 number is the sum of the two main male and female primordial archetypes - ADAM and EVE equaling 66 - is an indication that the balancing of the 6 natures along the male/female polarity is the goal of the religious path on one level. The primordial gender symbols actually are not speaking about two separate beings - Adam and Eve - but the two aspects of the polarity in every human being. That is the meaning and reason of interpreting the scriptures holistically and mystically; the allegory is referring to one being's psychology and soul-structure, not merely a tale or historical fable of a couple.

It is interesting to note that the Quran has 6666 verses in it according to some mystic sources.

One arrangement of 66 is God in the microcosm, and the other is God in the macrocosm 66.

As above, so below!

Explanation of Natures:

Nature 1. The Divine nature:

Feminine: Feeling / Masculine: Energy

This is the nature that is the source of all things. What is this source? It is spirit, which is energy, and also feeling (consciousness), which is essence.
Therefore, the source of all being is energy, and feeling.
Energy is spirit.
Essence is feeling. That describes nature number 1.
When people say spirit, something we hear all the time, they are talking about the building block of being, which is *energy*.
Intelligent energy is "Intelligent Spirit" a human being or what is called a "holy spirit" or embodied energy. Essence is what gives meaning to energy, or *life to it*. Life is feeling, feeling is essence, *energy is life*.
First Cause: When the Rau (disembodied energy) merged with the Tao (form) the first feeling was made, the essence began. That is the first cause, the big bang, and everything else that matters for

humans, from that, everything else began, evolving to perfection, then completion.

Nature 2: The Angelic nature:

Feminine: Light / Masculine: Power

The nature that controls all reality by light and power. This nature allows any of us to exercise power over anything as long as we have the circumstance of that power, or the intelligence of that power. Intelligence is light; circumstance is power over something. For example, even a baby can exercise power over something, if he is put in a position to, such as given a doll by its parents (Although he hardly can exercise that power by light, yet). Equally, the aspect of this nature wields power subtly by using intelligence (LIGHT), and always will work if the intelligence is based on truth.

Angels as embodiments are beings of light who adhere to nothing but the truth.

Now an ordinary human cannot (even though the angelic nature resides in us) be an angelic force entirely because he or she cannot adhere constantly to truth or light, we waver too much. However, an angelic force has no choice but to determine reality by light, and truth by power, OR POWER BY TRUTH.

Nature 3. Higher Adamic nature:

Feminine: Cause / Masculine: Effect

This nature exhibits the fact that reality exists in or operates in a science or knowledge - cause and effect. For example, it is a scientific process that creates us, that kills us, that heals us. Nothing is out of a vacuum; it must operate in a science in order to be; this reality is the essence of the third nature that I call - *Higher Adam.* No one dies of nothing; a pathology (science-knowledge) kills you. Everything comes and goes from us in a science or knowledge.

Nature 4: The Lower Adamic nature:

Feminine: Stability and Maturity / Masculine: Growth and Change

This nature is the determiner. It determines the balance of what we want. In other words if we are in a car and we are cold and turn on the heat, it is the *lower Adamic* nature that determines how much heat is enough to make us comfortable. This same Adam fell from paradise. Consequently, this informs us, this nature was responsible for our fall from grace, as the psycho-spiritual chronicles of the Bible and Quran states. This is the nature that is our most vulnerable point. Why is this the case? Because what this nature does is determine what makes us comfortable, balanced, and at the same time can cause us big trouble by turning comfort into pleasure, or *extending* it (Jinn).

Here is an illustration of this:

Imagine you are driving in a car to a destination, just driving, not doing anything else. Everything is normal as you drive, but all of a sudden, you realize you are a little chilly. Therefore, you turn on the heat and your lower Adamic nature tells you how much heat is good for you at the time. Without this nature, one would have no criteria for balance, comfort, and ease. One might freeze or roast to death.

As you drive along further you realize you are bored; so you look to entertain yourself by turning on the radio. You turn on the radio to some music. Now you are not only driving to a destination, you are also using heat to make yourself comfortable, as well as playing music on the radio to entertain yourself.

Driving a little farther, you realize you are hungry so you take out a sandwich from your lunch box and start eating, as you are still driving.

Now our little ADAM may be in trouble if he does not begin to control himself.

He may also start singing, whistling, using his cell phone,

really have a ball, then start smoking, putting his hand on his wife's leg, and eventually he could easily kill himself driving to work, because his extension of being (Jinn Nature) is overloaded. This is one level of interpretation of the allegory of Adam and Eve.
NOW WE CAN EXTRAPOLATE FROM THIS LEVEL WHAT HAPPENED IN THE GARDEN OF EDEN, CAN'T WE?

Nature 5: Jinn Nature:

Feminine: Stillness / Masculine: Extension

The extension of being, in any form. As mentioned above, the lower Adam extending from comfort to pleasure. This is why it is written that it was the jinn nature that enticed us to fall.

Nature 6: Anima Nature:

Feminine: Nurturing / Masculine: Consuming

This nature is all about maintaining the unit intelligent being. That is why animals are mostly always eating, or looking for food, that is the essence of this nature, not to eat, but to maintain, nurture, and survive the unit physical being.

Nature 7: Holistic self:

(Soul-Atman-MIND) it is the nature of the holistic self. This nature contains the other 6 and its purpose is to maintain the balance of the 6 inside itself and at that, reaches its completion.

The key to a successful life is to understand that these natures are balanced along the yin-yang matrix.

The imbalance of the 6 natures along our male/female inward polarity is the reason why the highly evolved beings who have charge of this affair of the ellipse have revealed scriptures, inspirations, words, religions, and mystic paths to humanity. These highly evolved beings include archangels, primordial Adams (our Fathers in Heaven) highly advanced and evolved female essence-adepts (Mothers in Heaven: Virgin Mary; Mary Magdalene, Isis, Diana, Kali) deities of immense power that taken together all equal what we can term essentially - God. God, not in the comprehensive, transcendent sense, or sky-God (do everything for us, like Santa Claus) but the only sense that matters; for the fallen human being, the only thing that is pertinent to our condition, is the clear explanation of what happened to us, why it happened, and how to get back our pure nature that we have lost. That is why ALL the symbols, allegories, scriptures (less those corrupted, misinterpreted, and misunderstood by the most profane and ignorant) point to this and this only.

And the two key structures in this fall and eventual imbalance are the celestial essence circles that have produced the ellipse on one level, and on another level have produced the imbalance of our inner transcendent natures, that all of the divine guidance and light that has communicated with us is trying to rectify.

When this is accomplished, all humans will return to perfection, and beyond, within the mysterious seventh nature, personified by the reality of the holistic self, in which preside the balanced, smooth operating machine of the 6 natures of being, balanced fully along the yin-yang or male-female polarity.

Because of the ellipse, our six natures are in an extreme state of imbalance along this yin/yang matrix of our being. All religion and other more pedestrian sciences such as psychology, sociology, as well as philosophy, are only trying with their methodologies to re-balance this inner universal phenomenon. As of yet these attempts have been highly unsuccessful, with the exception of a few people in history. This guidance is spoken about in The Quran:

The Six (7) Natures

"Then Adam received (religion, science, philosophy) words from his lord, and turned to him. Surely he is often returning to mercy."

Quran 2: 37

On this hierarchy of being, where the six natures primarily exist, the condition of the ellipse is reflected through the imbalanced state of the six natures. We, in our religious, and mystical methodologies directly impact the six natures (attempting to balance them) that then attempt to impact the subtle domain where the circles that constitute the ellipse and perfect circle exist. All this in an attempt to return the soul to the harmonious inner perfect circle, and even beyond this condition to the higher stage, where it's impossible for an ellipse to exist: that stage is the stage of completion.

Another way of understanding this is to say that when the six natures are rebalanced (however that is done) then the ellipse will no longer be. As the opposite is true, when the ellipse is resolved then six natures will automatically be balanced.

And that is and has been the goal of all mystical, religious methodologies since the beginning of our fall.

God, or what we call Evolutionary Intelligence (the first person designation of the cosmic science that rebalances man), that is the religious, mystical, scientific, philosophical guidance that has come to us from the standpoint of this fall, exists solely to rectify this.

YIN YANG

Yin Yang, the balance of opposites, is not really a balance of opposites, but a balance of energies. They are only opposites relative to their differences. Therefore, the reality behind this concept is about creation arriving at balance through the varying energies coming together in forming a whole that is trying to be functional. At one point, the holon is held together by the symmetry of Yin Yang; at completion, the transformation of the human sentient holon to the completed holistic self (soul) the Yin Yang of the six natures become hermaphrodite, or perfectly balanced producing harmony.

Yin Yang is not strictly a transcendent law. There are no innate opposites in reality, just different energy formations merged into a whole that transcends each of the energies - that is what creates Yin Yang. Yin Yang appears to be so prevalent only because we are at a certain level of being in the ladder of creation. The lower we go in that ladder the more we will see this mirage of opposites. This is because creation is only a mixing and merging of pre-existing energies, that at their highest end are divine (source) and their lowest end are filled with the conflict of Yin Yang. Yin yang only comes into reality at the mixing of these energies. It is the Yin Yang that holds them together.

There is only one transcendent essence out of 32 that can directly involve creation, and that is the "forbidden tree" of Adam. This tree is the only transcendent essence that can create anything. It is our proverbial 4th circle - the creative matrix of reality - in our Garden of Eden. This is the tree man has so pitifully stumbled into and become trapped in through the creation of the ellipse. Recall the fourth tree in a non-elliptical state is tremendously sublime, as all God's essences are.

All the evolutionists must understand that this powerful tree is the only essence-world that evolution, the condition of change, can exist in. Consequently, this informs us that evolution is not a transcendent reality in those other worlds, save through their own fourth tree. Therefore, in that sense it does (evolution) have a degree of transcendence - though not the physical evolution modern science is researching. This also informs us that the Hierophant most of the time, in abiding in the lawful descent (non-elliptical) in the fourth tree of creation, mostly enters it through psychic powers, not his physical being; even in far descents in the vine of this consciousness tree. We now know that this is the only method of entering this tree from the 5-source circle, lest one create another ellipse, and spend another millennium in "damnation."

At the genesis of creation, the mixing of these elements through the fourth tree precipitates the activation of Yin Yang. In the other eighteen transcendent unified essences, there is no Yin Yang, because all the essences are indivisible. They only can become divided, therefore mixed (creative potential) in a descent in the fourth tree. Therefore, we can easily deduce from these criteria, that Yin Yang only exist as potential in the fourth tree when its creative capacity is activated; therefore, Yin Yang is a relative phenomenon, not absolute. Moreover, it is clear that "creation" is a phenomenon that exists only in the fourth essence, as spoken above, that this is the circle of creation, and consequently destruction.

The specter of evolution is also a by-product of the creative matrix of the 4-circle essence. No other essences on the Garden of consciousness have in them a creative matrix (save in their own fourth trees); therefore they lack any kind of evolutionary phenomena. This can be clearly distinguished from the 4-circle world of destruction and creation, in that any world that exists outside of the fourth circle does not exhibit evolutionary phenomena. Though it must be mentioned here that a skilled Hierophant can create a world that is free from the creative matrix of evolution by a skillful arrangement of his descent in the tree that produces evolution. The Torah, in calling the fourth circle essence, the tree of life, confirms all the above conclusively: **Then the LORD God said, "Behold, the man has become like one of Us, knowing good and evil; and now, he might stretch out his hand, and take also from the tree of life, and eat, and live forever" - Genesis 3:22.** It is only in the tree of life where death exists!

Chapter 11

The Soul

The word spiritual means the condition of processing and change of energy for the transmutation of the human being.

All spirit is energy, and all energy is spirit. In addition, all that exists is energy. God = everything. Energy is life, existence, light, radiation; it is the essential reality of being.

Of course, there is *energy*, and there is energy.

Intelligent energy, or embodied energy is the Holy Spirit, which is the sustainer of the essence, and the part of that soul that stands between the essence (circles) and the Tao, which is the mediator of the soul with the Rau or disembodied energy, or spirit. The Tao is that which brings or filters this disembodied Rau energy to the soul, when needed for its sustenance. The mystic Lau-Tzu, says of this relationship between the Tao, and Rau: The Tao known is not the eternal Tao. The Rau is the eternal Tao - the free energy that pervades all being. Saying it is eternal is saying it is disembodied, beyond form.

The structure of our soul in a non-elliptical condition is a fluent running energy "machine" that is the most sublime process in existence and the quintessence of the divine.

The three structures are:
The Essence that has 22 elements
The Holy Spirit that has 9 elements
The Tao that is one *(Figure 20)*

This equals the 32 aspects of consciousness.

The Soul

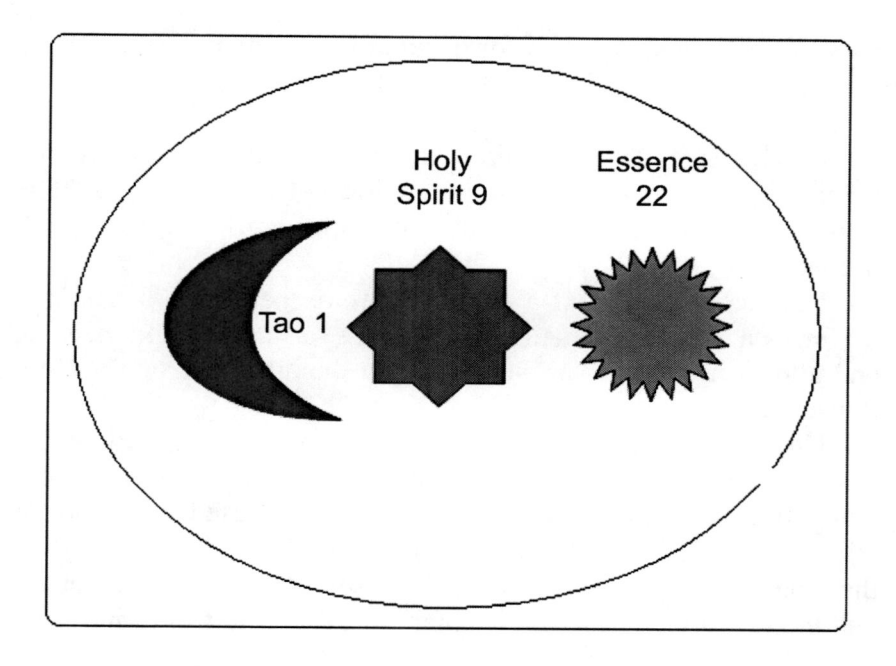

Figure 20: Replica of the three elements of the soul.

ESSENCE

The face of the soul is the essence, this essence is the primary aspect of our consciousness that we abide in all the time. It gathers its power and strength from the Holy Spirit, where reside the attributes and channels of energy that feed and nourish this essence. The often-recited metaphysical aphorism: 7 levels, 4 worlds, and infinite dimensions - the infinite dimensions are the essence, or the possibility of infinite dimensions comes from the essence. That is, though, only a possibility when the essence is in a non-elliptical condition. Of course, as the subject of this book enumerates, the essence is the ellipse (imperfect circle) humans have fallen to. Therefore, it is the essence that has been temporarily damaged. The Holy Spirit and the Tao do not have any damage to their structure or operation as individual entities. They are affected because if the essence is damaged then the entire soul is affected.

This essence as I have said before is the primary structure that manifests ALL of our states and stations of consciousness. It is also, as said above, the part of us that has been affected by the fall of man.

The Holy Spirit is of a different nature than the essence, and performs a different function. You might say it fuels the essence, and allows it to have various states via the attributes of the Holy Spirit.

This essence also is the "garden" of paradise related in the Western scriptures, Quran and Bible, as the allegory states, from which humans in primordial times "fell" to a debased condition for "disobeying" God. This is the reason why we are in a condition that needs religion, mysticism, philosophy, and other thought culture to bring us to the original state we fell from, and then beyond to a permanent station.

As for the human mind, it exists as a perversion of energy, deep inside the essence.

Because our soul is ellipsed or corrupted, there is a perversion of energy inside it that deludes us that we have a stable mind. This is false: it is a disturbance in the flow of energy in our essence that gives us the delusion that there is something like a stable mind in humans.

HOLY SPIRIT

The Holy Spirit is our inner intelligent energy structure that feeds and nourishes the essence, and brings in new energy from the Rau (eternal disembodied formless energy-spirit) through the Tao, in a recurrent cycle of eternal energy processing. The energy from the Tao is then processed and purified as needed by the Holy Spirit, before it reaches the essence. Therefore, the Holy Spirit is essentially a highly intelligent attributive power, as well as powerful and intelligent channel and processor of energy.

The Holy Spirit is where the attributes of God reside, and it supports the essence in its states and stations of mind.

What has been known in spiritual parlance as the Kabbalah reflects the structure of the Holy Spirit, albeit with a few differences. The Kabbalah has 10 elements; the actual Holy Spirit has 9.

The Holy Spirit fuels the essence and the Tao mediates or brings in new energy to the soul from the boundless spirit of formless energy - Rau, as needed by the soul-machine.

TAO

The Tao's function is its relationship with the Rau, and Holy Spirit, in bringing in and filtering Rau energy to the Holy Spirit.

One could imagine the problem of the soul in that one aspect of it, the essence, is dead because of the ellipse. The basic output mechanism of this energy machine is disrupted. One might imagine that the Holy Spirit and Tao therefore are inactive. That is not true; they have, because of the ellipse, made themselves into the guidance (God) channel which has assisted in producing the spiritual guidance that has come to man, in order to heal the ellipse. There is a modicum of energy cycling that goes on naturally between these elements that always, because of the ellipse, not only causes our death but brings us back to this matrix in order to resolve the ellipse perversion so the soul machine can return to its optimal condition. In other words, energy cycling still takes place but it is imbalanced, in that the essence only outputs negative energy to us, because of its elliptical condition.

THE SOUL IN MYSTICAL COSMOLOGY

<u>Christianity</u>
Father
Son

Holy Spirit

<u>Yoga</u>
<u>Brahman</u>
<u>Vishnu</u>
<u>Shiva</u>

<u>Islam</u>
The Lord of men
King of men
God of men

<u>Theosophy</u>
The Adept
The Goddesses
The Master

<u>Buddhism</u>
The Buddha
The Sangha
The Dharma

The three structures of the soul are represented in trinities across the entire spectrum of spiritual cosmologies. All these trinities are reflections of this inner world of the soul. The designation in Christian theology, of Father, Son, and Holy Spirit is, for example, not applicable to "God" but the soul.

The trinity of Brahman, Vishnu, and Shiva is not a trinity related to the deity per se, it is a Hindu doctrine that is not necessarily related to a transcendent God, only as a triad of action: creator, destroyer and preserver that most likely the early Hindu mystics related this triad to the soul or Atman.

The Islamic, Theosophical, and Buddhist triads also can be easily deconstructed on a high level to relate to the soul structure. These trinities are all synchronistically put together signifying the three transcendent aspects of the soul, in the Tao, Holy Spirit and the Essence.

SPIRITUAL TECHNICAL TERM CORRESPONDENCES:

God and the Soul

Deity [God]: Islamic [Allah]   Hindu [Brahman]   Neo Platonic [Source]

Self [Soul]: Sufi [Insanul-Kamil]   Yogic [Atman]   Neo Platonic [Divine Mind]

God – Allah / Brahman / Source - is the same thing, the unbounded infinite energy that is the source of all things and that is in and around all things.

Allah is the Islamic designation of the deity. Brahman is the Hindu highest designation of their many deity attributes. The Source is the Theosophical designation of the Divine. These can all relate to the concept of the Rau, as unbounded Spirit, outside of the soul, but yet the source of it, and its ultimate sustainer.

The Islamic formula that the devout Muslim begins any action with is a perfect illustration of the relationship of these divine energies. Bismillah Hir Rahman hir Raheem (In the name of God the Beneficent the Merciful) is the formula that all Muslims are instructed to recite at the beginning of any action. But on a higher level it is done for the creation of form that reality uses to "create" an extension of itself, out of itself. It is also the line of communication to the Atman or soul through the *Rahim* (attribute) that stands for the Rau that divine energy that enters the Atman (soul) through the Tao to support the soul in eternity.

The clear, precise delineation of the soul structure included in the Circle theory in this book clears up the mass confusion of these mystical realities that have plagued religion and mysticism for years. Where theologians, clerics, and certain mystics have been ambiguous about the meanings of some of these terms, consequently, they have lost their true meaning, and any holistic understanding of them. The claim by some that these concepts are not important because the path is devotional and based on virtues, therefore these issues of spiritual cosmology are not important is a sorry excuse for ignorance and confusion. This, as well as the legitimate differences of opinion of commentators has led to the lost of knowledge of these metaphysical realities. Moreover, these terms have never been able to convey the power of these subtle energies; they are simply inadequate in describing the reality of the soul. This is unavoidable, since mere words have never, and never will be able to convey all the facts of cosmic reality that only experience can do. Nevertheless, as best the intellect can fathom, humans are entitled to the truth concerning these subtle realities that rule them.

The paucity of the general understanding of metaphysics aside, the important thing here is seeing that all the paths use spiritual technical terms that refer to the same thing in each school of thought, just using a different language.

PERFECT CIRCLE VS. THE ELLIPSE

There are two levels of the perfect circle or soul. One that exists before resurrection and the one that exists after, something I will describe more of later. Before resurrection describes perfection, after describes completion. We will have more on this in chapters 14 and 15.

The Tao is the element of the soul that brings in energy from the eternal Rau, as the organ of the soul needs it. The Tao works directly with the Holy Spirit that in turn nourishes the essence. This is a smooth running consciousness energy station/state machine that is absolutely perfect. As stated above, though, the ellipse is a broken soul machine, and the energy configuration and operation exhibited in the smooth running perfect circle described above does not exist in the ellipse. That is the reason we suffer - the energy flow is blocked and distorted, and the operation of the 3 elements are stifled, in that humans are trapped in a lineage or vine, or tree of the essence, which is the imbalanced ellipse structure.

Normally, in the perfect circle station-state activity - when one is traveling in the lineages, vines, or trees of the Garden - it operates smoothly, and sublimely. On the other hand, in the imperfect circle (ellipse), the capacity to travel in the station-state paradise is not even operable. All we get is a trace of feeling from it. This makes all humans experience-*only* energy warping, that makes our existence one of suffering. In essence, the machine is broken; therefore, we have the life we have in this world. When we die (because of the ellipse), we are recycled by the energy and intelligence of the Rau, Tao and Holy Spirit to try to attempt again to return to the perfect circle by defeating the ellipse. This is the essence of what Buddha taught in the doctrine of birth and death that constantly keeps us in a cycle of rebirth in the ellipse until in one of our lives we come out of it, usually through effort and metaphysical science.

As for Western metaphysics and its concern with judgment, the end of the world, and resurrection, that is entirely about the time-centered macrocosmic era that deals with the dissolution of the ellipse; as well as expansion of the soul. This begins the heaven and hell period on earth, which is described in chapter 13. First, we are going on somewhat of a diversion, in that we will cover the fact of God in the next chapter, though one must confess that this book all inclusively is nothing but about God.

Chapter 12

The Divine Epiphany

In order to understand God one must consider certain criteria and certain transcendent facts.

Criteria
God is one or a unity.
God is unique in developmental phenomena.
God is everything, in everything, and is made of everything, and everything is made of God.
There is nothing that is not God.
There is nothing separate from God.
God is not a personality or being.
God cannot be confined to one thing.
The divine includes truths as transcendent facts.
God has a passive aspect, and an active aspect.
God does not create anything out of nothing:
He creates by weaving together aspects of himself.

FACTS OF GOD

The fact of the reality of being, defined by the 13th century Sufi mystic, Ibn Arabi, as the unity of being. This is the fact of unity; all are interconnected in a non-dual state. In this sense, everything is God. To exclude anything from being God is a form of idolatry. This is not pantheism in the strict sense, since this idea of the divine does include the recognition of degrees of being.

This goes to the point that one cannot confine God to anything, particularly in any anthropomorphic sense; hence, he or she is not exclusively existent in any being. The God state the lofty epiphany of developmental evolution, is for all. Nor can one confine God to any phenomenon, for he is above it, as well as in the phenomenon at the same time, yet that is never what he is confined to exclusively. In other words, God literally is all things, yet is figuratively unique when we enter the field of development. Then he/she becomes discretionary, because we are traveling from one point to the next, which requires identification with form. In the state of enlightenment, the inner eye sees through these relative forms by seeing the absolute in and beyond all form.

On this metaphorical level, which relates to developmental phenomena (see underneath) God is unique in the sense of being the best of anything, or the conclusion of the development. Therefore, we can say that to arrive at the highest spiritual stage is God - the divine station-state. This reference to God is the basis of Western religion and mysticism - the God of degrees. In the best sense, this defines Western metaphysics as a science of levels. However, this science has gone astray by taking the doctrine too literally, thereby eschewing the other aspects of God.

The confusion by exoteric religious scholars of the being of God, his "creations" with he, the entity God, is in their false notion that God is literally a single entity of some sort. Ironically as they accuse others of idolatry, it is in fact they who are the true idolaters - in singling out some ethereal super being as being God exclusively, a being they or no one in history has seen, or identified on any level. In fact, this fantasy being has more to do with the comic strip hero Superman or Santa Claus, than any reality.

There is the state/station of God in the developmental phenomenon of being: holons, and the holistic self (God in development), and the existential reality of the facts of God, or his infinite eternal systems of himself: worlds, levels, and dimensions.

Our next fact is dharma or cosmic law, the natural existential law of reality. God did not create cosmic law, he is the cosmic law; it is his nature. This aspect of the divine was most emphasized by Buddha. He taught that the key to spiritual success is yielding to the universal cosmic law through integration on all levels. This is similarly taught in Taoism. One should integrate their will with the will of the intent of the cosmos. This fact is the basis of all genuine metaphysics of an esoteric mode.

All religious systems aim at integrating with dharma. This is based on the premise that somewhere along the road of man's evolution he became detached from his original nature, thereby losing his relationship with dharma. Evolutionary mystical theories that do not accept the idea of the fall of man, nevertheless usually accept this fact of dharma. The correct application and understanding of the body-mind-heart practice of a genuine metaphysical methodology can merge the wayward soul back into union with the cosmic law.

Another misconception that exoteric religions have fostered is the belief that dharma has favorites. (This erroneous concept though is primarily held by sky-God believers who imagine the divine favors their particular cult-creed from amongst many). The exoteric religions have this concept so confused that they will always in the end fight and divide over which sky-God loves whom the most. This is obviously a symptom of ethnocentric consciousness disguised as religion. Dharma ONLY favors those who remain integrated with it, for dharma has no personality, as the other aspects of the divine epiphany are without personality or favorites as well. In other words, we are on our own; there is no savior, intercession, or any personality going around saving or helping people in this regard. However, this does not in the least preclude another being from helping another, to a certain degree, though it is the individual who has to take the medicine not the helper. In addition, the various teaching archetypes, such as Jesus, Muhammad, Krishna, and Buddha are usually always available in the world to serve as a guide to spiritual success for sincere seekers. In fact, these elements are latent in all humans.

The next fact is the reality of levels of being: levels, worlds, and dimensions. These are stages of development that are guides to change, that involve the transformation of energy. This supports the uniqueness of God, indicating the best of anything, or the highest level, which is indicative of completion. It also indicates balance, which is an indication of perfection. Again, here is where the exoteric religionists go astray, in not understanding this developmental aspect of the divine epiphany, and supporting a false notion of an antiseptic sky-God far removed from his creation. This aspect of God does have an exalted stage that is a metaphor in the sense of the highest degree of the divine state of being, the intent of the Evolutionary Intelligent God. A level is something that resides in a being, or world. Levels do not exist in space on their own they are relative to living developing life, or being. A world is a place where the frequency of light in a confined area of space is universal; therefore, it is a unique composite. A solar system for example is a world. It is also a level, in the sense of being a part of the galaxy. On the other hand, a dimension is a unique world based on its dimensional frequency (not its level) within a holarchy of being. A dimension exists on a body or plain of other worlds that are related, but have different dimensional frequencies, thereby cannot be seen by each other interiorly, they can only see each other exteriorly. For instance, if one is on a holarchy of worlds that include worlds a, b, and c, then on the plane (level) of the holarchy they can see each other, but once one enters for example, world a, then one will not see world, b, or c in their original form, nor can they see a, at this point. The essence, tree, Garden of Eden, or circle, the subject of this book is a world that exists as a level of the soul, whose dimensional reality is unlimited.

    The next fact is the fact of what is called the holon. Any sentient or non-sentient entity is a holon. All holons, from humans to sub-atomic particles, all the way to distant stars, are reflections of God from the highest world to the lowest - which is the material state they mirror God. In other words, all holons have properties in the higher realms, though they - as the pole experiment below illustrates - are not always in tune with their higher realm. The more developed a holon the more in-tune it will be with its higher reality.

The Ellipse: The Fall and Rise of the Human Soul

All sentient holons have to have a matrix to exist; mostly, non-sentient holons do not require a matrix to exist, though they are part of a holarchy. Yet they do make up the physicality of all sentient beings. Non-sentient holons can exist - and do make up the infinite material of God - independently from matrixes; in other words, they can exist independently, though not independent from the infinite spirit called the Rau, which is unbounded spirit, our ULTIMATE fact.

Holons have an aspect of the four worlds, six natures, seven levels, four elements in them by existential reality. In other words, all who are "created" will have these passive aspects of God in them. Actively they will exhibit attributes as holons, and by the dictate of the unity of being, may evolve to the highest holonic state - the holistic self, as regards sentient holons.

All holons are "created" by their reflection of a reality in a higher realm. This is based on cosmic law revealed by Hermes: "AS ABOVE, SO BELOW". The below is never an exact replica of the higher world. For each realm reflects reality based on its own frequency of light. No holon is a special creation of God, including humankind, as if God one day decides: I want to create a man, or a dog, or a cat, or a star, or solar system. This is fantasy, and unreality. Creation is a complicated process of inter-dimensional holarchy that involves chains of being that stretch from the highest world to the lowest of God's passive systems of being. The holons always are part of the systems of God, not separate from them, as he God is not separate from holons nor his systems.

POLE EXPERIMENT

This can be understood by imaging a pole being placed in the matrix of the four worlds, and visualizing what one would see in each world of the pole. Each visualization of the part of the pole that appears in each world will be different. This is because the worlds have varying degrees of frequency and

other different elements. The pole is not only appearing to be a different phenomenon in each world, it is a different phenomena, but yet the same. It is only different because the mirror in each world is different. A world is a perspective and a consciousness just like our different perspectives and consciousness as sentient, developing holons.

The question arises of how can one see this pole going through these different realms, displaying apparently different objectifications of itself as the single entity that it in reality is, and not 4 different facts? The answer to this is that one has to see the reality of the first fact, the fact of the unity of being to ever visualize this pole as the unity it in fact really is.

Another question is. Who is seeing this pole? Most people are seeing the pole only in the two lower worlds of mind and body, but since we are disconnected cognitively from the two higher realms of spirit and soul, we will not see the pole in those realms.

We can deduce, through scientific capacities, that the pole is invisible in the psychic realm, and visible in the material realm to our physical eye. This of course we know is illusionary, since anything and everything has its degree of physicality, it is just that we in this material world cannot see everything by the lens of the physical eye.

The next fact is the reality of the holistic self, and the six natures. I add here the six natures because the holistic self is the 7th nature, the highest holon. Beyond this, there is no more development, just manifestation of being.

This includes the fact of the soul, or God inside the self. Beyond it is only the transcendent reality of infinite energy - Rau, which is not a holon. The 32 infinite levels of being exist in all souls: latent in undeveloped holons, and apparent in developed holons, or holistic selves. All complete souls or beings (holistic selves) have access and consciousness of these 32 levels.

These 32 levels as entities are indivisible; they cannot be divided, or separated. Though they exist in three different levels of the soul. Through the essence they can be extended ad infinitum. This essence and its potential for unlimited states of consciousness, our Garden of Eden, the subject of our treatise, is the meaning of infinite dimensions in the aphorism: 4 worlds, 7 levels, and infinite dimensions.

When God "creates" something, he first establishes the intention in the highest world, and then the intention is reflected in the world underneath him - the world of the domination, which rules all holonic structures, by light and power. Here the "creation" is designed. Then onto the next lower realm of the psychic powers to implement and operate the design from the upper realm. Lastly, the fact of this creation comes into being in what is called the kingdom - the matrix or environment of being. It is the matrix that is the kingdom, that which lives in it is the holon, or holistic self.

All reality is the Rau-unbounded spirit, plus holons, and the holistic self or completed self, the highest holon. That's it. There is nothing else. Though, within these varying worlds, realms, and the 32 levels of being are an infinite variety of consciousness, levels, dimensions, states, and stations in this arrangement.

The Rau feeds the divine soul, or Atman, through the eternal essence-spirit known as the unique-Tao. This Tao is the only essence that relates to the Rau. The Tao in turn develops and nourishes the Holy Spirit. The Holy Spirit then nourishes the third part of the soul we know as the essence. These bodies of light in the soul are the 32 levels of consciousness of Buddhist cosmology. There is nothing other than these 32 levels of transcendent consciousness-being that make up what we have been calling the soul, plus the Rau, God's systems, and ascending, and descending holons, all are ruled by the dharma save the Rau.

Here again is another confused idea by the exoteric literalists who have mistaken the exalted or solitary position of the Rau, and twisted this concept into a concept of a distant God that doesn't dare dirty his holy hands with the other aspects of himself, thereby they label these aspects "creation," in fact a misapplied word that can't apply to his systems.

The Rau cannot be confined by anything. It surrounds, and is in, outside of, and inside of, everything and all, space, non-space; in fact, it is always everywhere at the same time!

There is a Sufi tradition of Muhammad asking God whether something can contain him. God answered that nothing can contain him, but the heart of my worshiper. This is a reference to the soul, as God's representative active being through his 32 indivisible spirit essences that reside in the soul. The Rau though cannot be numbered; therefore, that aspect of God cannot be contained. If one makes a statement that "over there is God" referring to a holonic being, sentient or not, or some kind of phenomenon as being God, then at best this can only be symbolically true, in the sense of developmental phenomena, as regards God being symbolically the highest stage in a matrix of development. But that's as far as it goes, for the statement is essentially a false one. This is proved by the story of Moses asking God whether he can see him. The Quranic tale goes on to record God saying to Moses "You can't see me. Though look at the mountain, if it remains firm, then you will see me. The mountain dissolved before Moses' eyes." This clearly can be interpreted that nothing can exclusively contain the divine to the exclusion of other things. Therefore the idea of anyone ever seeing God in any literal form is an impossibility, since everything around the site of God would have to be annihilated as the mountain was in the Moses story.

Taking the above in summation, we can conclude that God is what I term truths: Haqiqa in Sufi parlance, which means reality being, reality truths, reality waves, and reality systems. These systems are NOT created, they are God's nature.

God has a passive existential reality and an active existential reality. The passive aspect of God is the imprints that make up all aspects of the active manifestation of God. For example, the Rau, unbounded spirit underlines ALL reality, passive and active, yet it is a passive reality. Of course these terms, passive and active are relative. This formulation can be understood by providing examples of its operation.

For example, all holons have aspects of the four worlds, six natures, seven levels, four elements in them by existential reality. These passive aspects of reality are not created; they are the infinite patterns of reality. In other words all who partake of being will have these passive aspects of God in them. Actively they will exhibit attributes as holons, and by the dictate of the unity of being, are evolving to the highest holonic state of the holistic self-completed soul. It is dharma that mediates these varying forces, as well as rules them, when they become active.

EXISTENTIAL DIVINE

Passive

Infinite
Rau
Dharma
Unity of being
Natures
Worlds
Elements

QUINTESSENTIAL DIVINE

Active

Eternity
Atman (soul) rules active aspect of God
Attributes
Holons

GOD AND CREATION

The question arises, something we have touched on briefly in other chapters: what is God and what is creation? It may sound contradictory by implying that everything is God, and also acknowledging the fact of creation. Well, first of all, the implication that everything is God is true on the level of our first fact: that is the unity of being. It is not completely true when we deal with the fact of God's developmental systems. This is where creation appears. First, let us digress to develop the real meaning of the concept of the divine or God.

This concept of God is essentially beyond the ideas of omnipotence usually associated with the concept that defines sources of being. As we said earlier, the primary sources of being are the 32 levels of consciousness-being in the soul. These levels are uncreated, that is why they are divine in a literal sense. They are the great source of all "created" reality that is in fact not created in the sense we ordinarily understand by creation.

We think of creation in the same way as we think of a magician: as if God creates something from nothing. This notion is in fact absolutely false. In fact, literally speaking, God has never created anything, since all substances, being a part of him, have always been here and always will be here. If the question arises, where is here? Here is life, and consciousness being.

Literally speaking, God creates as we would create a car, for instance, from existing materials. He, like we, weaves different substantial materials of himself, to manifest a novel manifestation of parts of himself. The intelligent systems of creation, some he has created as grid patterns, through sound phenomena, are the weavers, or the programs God uses to manifest himself in secondary active and passive elements of himself. As described earlier, this can only be done in the fourth tree of The Garden of Eden, where the existential reality of the creative matrix exists. This fourth tree, of course, is one of the 32 source essences of reality, as we have said earlier. When these creations - which are never sentient beings - are finished with their prescribed purpose, they die, and the materials of which they're constituted return to their original source. All these materials are infinite, the idea of death only relates to the creational aspect of being, that has a purposeful existence. This creation, in fact, is a corruption of reality, in that reality makes something that is created from infinite source materials that become subject to the secondary entity, which is the result of mixing the source materials for a temporary phenomenon that appears to transcend the original infinite substances of reality. This is an illusion, for the infinite source materials never die; it is the temporary creation that is truly mortal.

Can anything created therefore live forever? Technically, no: the very nature of being "created" automatically involves inevitable death. This is so because ALL creation exists in the 4-circle of the essence, the creative matrix of reality, which is also the matrix of death and destruction; outside of that, nothing ever tastes death, because it was never created in the first place! Please keep in mind that this creative and destructive matrix IS THE WORLD WE HAVE BECOME TRAPPED INSIDE! This is the very essence of our story here, about the ellipse that has befallen our soul for a time, and allowed us to taste of the tree of life and death, until we return to the tree of light upon light, that never, ever goes out!

Chapter 13

Paradise and Hell

What are the major implications of the knowledge revealed in this book? Simply put, there will be - or it already has begun - a unity of celestial forces that will come together to rectify the world of the macrocosm. Consequent to that will be the involuntary rectification of individual humans on a grand scale. This will not be a pleasant experience. The macrocosm in purifying itself and returning its circle to its original perfection is the sole reason for the inner and outer turbulence the vast majority of humankind will experience in this period.

The literal form these events will take in the panorama of our life experience, for the most part, will not be anything we will expect. The forces will employ unique as well as familiar apparatuses and actions to fulfill this mission to rectify the macrocosmic circles, and purify them of negative energy. Moreover, since humans are a vital part of this milieu, consequently they are inextricably involved with this purification.

In this period, everything in the cosmic order will come together and bind all humanity, with no possibility of escape. These celestial forces will have indirect control of humans, via their control of the macrocosm, and direct control of humans because the assignment of fulfilling the transformation of energy (chapter 5) of the hidden fourth world is still an active reality. On one level, they will not be interested in affecting humankind, though this level is not peripheral, in that on other levels the direct control of human beings is relevant to their work. Any idea of humans avoiding this is impossible. This means that they have direct access and rights over any human alive or dead! In fact, the intelligence's omnipotence over humans will be so apparent that one will clearly see that even if one did escape, that would only delay the inevitable. All have to be purified, by order of the cosmic laws that are inescapable. One's lack of belief in this process will have no effect on the fact of these forces.

The good news is that this will not be a black-and-white scenario. A very few amongst humankind will be free from this. Those amongst humankind who, with sincere intent, have attempted to extricate themselves from the ellipse experience will have an easier time in this period of purification, based upon any progress they have made. This, though, will not be based on any protection for beliefs in any sky-God, but premised on real knowledge they have attained on the path.

What is the reality of hell? Many people say that hell is the condition of suffering that all humans have from time to time experienced. There is some general truth to these formulae, but in fact, the condition of hell as expressed in the psycho-spiritual chronicles in relating to the dissolution of this world-ellipse is another matter entirely. This hell is the period when the macrocosmic ellipse dissolves and rectifies its corruption. Simultaneously all microcosms will be naturally exposed to this rectification. Most humans will not be prepared for this, in enormous numbers.

Interestingly, as mentioned above, the Arabic word for hell - Jahanum - is related to the Arabic word for paradise, which is Jannah (garden-paradise), both derive from the root word, Jan (Jin) - to conceal. Concealed in the garden is eternity. As mentioned and demonstrated before, the ellipse-hell, Jahanum, is derived from Jannah-paradise or Garden of Eden. It is evident then, that hell is derived from paradise - ELLIPSE (IMPERFECT CIRCLE) FROM PERFECT CIRCLE. Therefore, we have an indication in these words of our earlier assertion that hell is a corruption or is a derivative of paradise.

Parenthetically, another fact here must be pointed out. The word "heaven" - a designation in some exoteric religions for paradise in the hereafter - is a misnomer. The dichotomy is paradise-hell, not heaven-hell. Heaven in metaphysical terms is a word that has to do with lofty thought and some forms of meditation. The opposite of heaven is earth, which would mean dense, earthly, or carnal thoughts, as opposed to open, heavenly thoughts.

All this relates to the time-centered event when the ellipse is to be dissolved in the macrocosm, bringing forth the scriptural paradise-hell epoch in this human dimension. It must be understood that the painful aspect of this experience relates to two aspects of the cosmic law that initiates this inevitable reality. One aspect of the cosmos relates to the cosmos itself. By cosmos I mean the world, garden, or ellipse, solar system itself, bringing the missing elements of its world back onto itself, eventually thereby reigniting itself to an operable stage. As I mentioned earlier, this involves the macrocosm regaining its 11 missing circles: the 4 lower, and the 7 angelic. Somewhat similarly, the microcosms will return their 9 missing circles: the 2 lower and the 7 upper angelic circles. We must recall these subtle differences in the scheme: the macrocosm has 11 missing circles. The microcosm (all individual humans) each has 9.

In the microcosm, there are two different ellipse beings sharing the 4 missing circles in the macrocosm: the macrocosm is missing the 2, 7, 1, 8 circles. Some people have been given the burden of missing the 2, 7 set, as others are missing the 1, 8 set.

This dissolution of the ellipse involves disruption of the solar system on a massive scale in the macrocosm, eventually affecting the microcosms.

The other aspect of hell that involves extreme discomfort is related to the original choice of the supreme intelligence to utilize the ellipse for energy processing in its period of being, as we mentioned in an earlier chapter. Unfortunately because the dissolution of the ellipse is not aligned to this energy processing, there is more of this processing that has to be done, therefore this will be reflected in the milieu of hell in this later period. This is related to the mysterious 4 world that lies underneath ours that the intelligence is transmuting, written about in an earlier chapter.

The opposite stage of this hell, those few who will be in paradise, are those who have before the initiation of the elliptic dissolution have, through their own efforts, already resolved their inward ellipse and returned to their perfect circle. The Quran quotes this archetype: "This is what was given to us before." In other words, the individuals who resolved their ellipse are now enjoying paradise even before the hereafter period!

This is an inward reality, not anything based on a gross earthly condition having to do with materialism, delights, or earthly pleasures. Certainly those in "paradise" will experience certain pleasures, particularly by comparison to those in hell, but this will be nothing arbitrary from any God, or sky-God, giving them anything from "heaven," as the understanding of heaven and hell from the exoteric religions have taught.

There is no certain information at this point of the length of the hell period. It is though impossible that it will be eternal. Besides the horror of the reality of heaven arbitrarily resigning any sentient being to eternal suffering - particularly in view of the certain cosmic reality that ultimately God, or heaven, is responsible for all things good and evil - cosmic science allows for hell only to be temporary for the individual as well as for the macrocosm. The intention of "hell" is primarily to purify, not to punish.

In fact, an ellipse being is technically already in hell. The hereafter period only enhances this condition, and also allows us to see this clearly, as the macrocosmic re-perfection brings forth the mechanics of our own re-perfection.

"AND OVER HELL ARE NINETEEN"

The science of the hereafter

This combination of cosmic law and divine dictate (in the processing of the hidden 4-world we abide in) is an awesome display of the divine will and cosmic science. Here is the list of this combination of forces.

By the order of time: that is the event that precipitates the last circle returning to this sphere - will initiate the combined celestial forces to begin their work.

The metaphysical dictum of heaven and earth uniting is the result of the ejected circles returning to the macrocosmic soul sphere. These phenomena will accelerate time, allowing the intelligences to correct imbalances instantly. The fact of our direct connection to the macrocosm (something hardly seen in ordinary reality-time) will be manifest because of this union of heaven and earth.

The law of association of energy, known as the karmic law, will aid in this process by obligating people to the intelligences that will preside over this affair. The active intelligences will not have an association with any religious or racial group at all. The concept of God favoring anyone has no basis in reality. Even those who have defeated their inner ellipse before this purification period have nothing to do with these events. The intelligences will ignore these individuals, as their energy does not require attention.

Nineteen of the circle intelligences will be physically active in this period, with the three-representing the core inner circles (5, 6, 15) being inert, or inactive. However, being the core of the ellipse - a circle that is to be purified - they have a unique interest in the outcome of the hereafter period, for it is their tree that is affected. Nevertheless, these three circles will maintain somewhat of a distance from the active work of the other nineteen circles. This will allow protection in the physical macrocosm for those who defeated their inner ellipse before this apocalyptic period.

In the cosmic paradise or perfect circle that produced the ellipse in the microcosm, then macrocosm, there are nineteen circles and a core circle with two inner circles, equaling 22 circles - that is the perfect circle before the creation of the corruptive circle or ellipse. Remember that the ellipse or imperfect circle is created when the fallen Adam loses consciousness of 9 of his circles and the operation of his perfect circle or his paradise. This occurred when he entered the forbidden "tree" or the 4-circle from the 5-source circle, whose energy is too powerful for the being from the standpoint of the source circle. At that point Adam lost the operation of his inner paradise because two of his lower circles were ejected from the orbit of rotation from his perfect circle, thereby making him lose consciousness of the higher 7 angelic circles. Now the numerology goes like this: 22 become 13: the two lower circles we lost and the seven angelic circles, this is in the microcosm.

Parenthetically, though, the numerology means little in that since our inner circuitry (rivers) were severed we have lost contact with our inner paradise and only a trace of our real selves can be perceived by us, or a mere shadow of even the 13 circles that our consciousness reflect.

Therefore, as in the macrocosm, as in all of our individual circular souls, we in time will gain access to our two ejected circles from each ellipse being, the 1, 8, and the 2, 7 ellipses. Then the microcosm will go from 13 to 15, re-perfecting the lower circles, and then our seven angelic circles come into sight, bringing us back to 22 circles. The macrocosmic reality at purification in the last days is slightly different. The four ejected lower circles return to the macrocosmic ellipse, thereby allowing the macrocosmic being to see the angelic circles again, (similarly to the microcosmic experience) and the 11 becomes 15 then 22, by seeing the newly-perceived seven archangel circles.

The cosmic friction in these phenomena will produce what we have come to understand as hell - which is a form of purification. These huge entity forces will begin the purification of these worlds or circles. The three inner circles, that is the 5, 6, and 15-circle, will not have to be purified, because they are the inner core or (axis) circles which were never really corrupted and the power and grace of them, despite the ellipse, allowed us enough light to be for a time in this corrupt world. The other 19 circles in the macrocosm will be purified and that will be the macrocosmic being purifying its world. "Over hell are nineteen."

At that time most of us will experience varying degrees of

discomfort or "hell", depending on our spiritual condition at the time of the great purification. On the other hand, those of us who excelled spiritually will abide in the 3 inner circles, sort of at the eye of the hurricane, where there is calm. This abiding in the three inner circles in this instance is an outward phenomenon. In other words, since they are purified inwardly, it is only the outward aspect where they could be vulnerable to this period's problems; therefore they will abide outwardly in the macrocosmic three inner circles. Those are the people in "heaven" or paradise who in fact, as the Quran says, will say in paradise: "This is what was given to us before." In other words they had already arrived at a degree of spiritual perfection before the challenging period of universal purification has arrived. This is the meaning of the statement in the Quran: "over hell are nineteen." That is, the universal cosmic entity forces over the nineteen circles to be purified from centuries of negative energy generated by the ellipse in our history.

This illustrates clearly that it is universal law that is the source of our return to perfection: ellipse, apocalypse, the words are "coincidently" similar.

ELLIPSE - APOC(ALYPSE)

The essential message of this book is that massive cataclysmic change is the inevitable lot of all humankind. This is the very meaning of the *apocalypse - dissolution of the ellipse.*

What do we ordinarily understand by the apocalypse? The end of the world, though the actual meaning of apocalypse is "to reveal." Since the ellipse period of purification depends on the world becoming aware of the four missing circles, and the seven unseen angelic circles, consequently the meaning of the apocalypse aligns perfectly with the events enumerated here. These hidden elements will be (revealed) to us, bringing forth a new world (returning to the perfect circle) and ending the world of the (imperfect circle) ellipse-*apoc(alypse)*.

It is inevitable, as all our Prophets, more so the Western

The Ellipse: The Fall and Rise of the Human Soul

Prophets, remind us, this ellipse (circle-world) will dissolve back on itself to the source Garden of Eden. This will manifest - Above - the macrocosm, as well as - Below - our individual selves, the microcosm.

Before the macrocosmic ellipse (being) dissolves this circle of consciousness back to the source-5-circle, it must first restore the circle (ellipse) by returning it to its original 4-circle perfection, or dissolve the ellipse. In other words, it cannot return to its source Mother Garden (5-circle) until it resolves the ellipse (4-circle, where the ellipse existed). Remember, the ellipse began at: 5 to (4=FORBIDDEN TREE) consequently, it is the 4-circle in which the purification is taking place.

When the dynamics of these phenomena begin to re-perfect the ellipse in macro, the great majority of the human race will be severely challenged, save those who have already resolved their individual ellipse!

This "hell" period is basically a form of purification, not any arbitrary punishment from an angry Santa Claus-like sky-God - an absurd notion. Moreover, it is not eternal, another absurd notion from the exoteric religions that have distorted this sublime science I describe that is based on precise spiritual laws of the universe.

In the opposite stages of this "hell" will be those few who will be in paradise or heaven, those who have, before beginning the elliptic dissolution, through their own efforts, already resolved their inward ellipse and returned to their inward perfect circle.

To outline these important cosmic phenomena further it must be said that in the microcosm, that is each of our personal beings, there are only two missing circles, 15 to 13, whether one is a 1, 8 or 2, 7, ELLIPSE. Alternatively, in the macrocosm there are four missing circles, 15 to 11; as if it were that the two distinct elliptical microcosms (2,7 and the 1,8) have split between them the burden of the macrocosm. There is no microcosmic being with four missing lower circles.

Microcosm 1
1, 8, ellipse: Originally fifteen lower circles plus seven higher angelic circles - 22:

Went from 15 to 13 at descent in 4-circle from the 5-source circle; consequently to the loss of control and consciousness over the seven angelic circles. Therefore, the number of essence-perfect circles goes from 22 to the elliptic condition of 13.

Microcosm 2
2, 7, ellipse: Originally fifteen lower circles plus seven higher angelic circles - 22:

Went from 15 to 13 at descent in 4-circle from the 5-source circle; consequently to a loss of control and consciousness over the seven angelic circles. Therefore, the number of the essence-perfect circle goes from 22 to the elliptic condition of 13.

Macrocosm
1, 8, 2, 7,

Originally 15 lower circles plus 7 higher angelic circles-22:

Goes from 15 to 11 to reflect a singular entity as losing 4 entire circles. NOT 2, as in microcosms. Consequently, the macrocosmic ellipse in history has had 4 movements of missing circles out of its sphere of influence. As a result of this, the angelic circles as well were hidden. Therefore, there are in the macrocosm four circles that will return. Wherever we are now in the order of return of these circles, on the return of the fourth and final circle will ensue the last days - THE END OF TIME.

The coming together of the above and the below in cosmic time is the accelerated science of justice that is inevitable. Actually that means a huge problem for us, in that the subtle workings of the cosmos will be accelerated where the above will quickly raise itself in the below. In other words, usually what happens in heaven takes a long and subtle process to reach us - and we hardly feel it. In the purification era, that will not be the case, phenomena are magnified. For example, as justice is usually slow in our present world, in the hereafter, because of the coming together of heaven and earth, justice, in a miraculous way, will be swift.

This is not anything arbitrary put into action by God or heaven; it is based on the laws of karma, and dharma. The classical exoteric religious understanding of heaven and hell is far removed from anything resembling reality. The thought of a sky-God disseminating succor to his believers, in some unscientific arbitrary manner, is so false as to be almost criminal to teach. Also, the absurd tradition of some horned demon running eternal hell to arbitrarily torment evildoers is equally false.

Paradise and hell are existential; they happen not arbitrarily, but by the cosmic laws of cause and effect - an aspect of karma, the law of time and place, as well as the law of dharma. Dharma is the law that underlines all cosmic laws. It is about one's knowledge of the way the universe works, and more importantly, one's acceptance of it. That acceptance, determined by knowledge of dharma and consequently doing positive acts accordingly, is the determining factor of whether one is truly accepting of the cosmic order.

Chapter 14

Return to Perfection

Understanding metaphysics and the path of return

"We said: "Get ye down all from here; and if, as is sure, there comes to you Guidance from me, whosoever follows My guidance, on them shall be no fear, nor shall they grieve."
                                                                Quran 2: 38

"But those who reject Faith and belie Our Signs, they shall be companions of the Fire; they shall abide therein."
                                                                Quran 2: 39

The guidance spoken of above is the religious, mystical paths, philosophy, and other sciences that have sprung from the guidance the EI has sent to humans through the angelic and transcendent Adamic nature. Few modern people accept the idea of return to God. Essentially, modern and postmodern man is a disbeliever in this cosmic affair of return to God or the lost perfect nature. On the other hand, if he does believe, it has degenerated to a crass exoteric distortion of its original intent. The original guidance from the Evolutionary Intelligence is hardly recognizable today. This degeneration of the religions themselves and the non-human practices of many religionists has, to put it simply, turned many people off totally from religion. Due to the overly ethnocentric, narrow, and selfish ideas of mainstream religion, they have precipitated the absolute degeneration of religion and become useless shells, and counterproductive to the very essence and meaning of spirituality. There are only a few esoteric paths today that have maintained a modicum of genuine spiritual guidance. As for the exoteric paths, there is none worthy to mention that retain any useful guidance.

Genuine spiritual practice starts with an understanding of the theoretical fact of what one is doing. That is called the exoteric path, or aspect of the outward.

The esoteric or inner reality of the path is the alchemy that occurs in the heart by spiritual practices based on the original - and updated theory - of metaphysics that always restores the guidance to its original purity. (This never lasts in the outward because of the nature of the fallen human. He continually distorts the messages from the divine.)

It is not that this path is split in two; it is that it has an outward part - theory, outer law - as well as an inward (esoteric) aspect to it. The unified path unites this system in a holistic sense, where both sides of the coin are given their proper due, thereby finding balance. Some paths, such as traditional religion, have totally neglected the esoteric, thereby rendering them useless as a mechanism out of the ellipse and on to enlightenment and completion. On the other hand, some paths have become overly reliant on the esoteric, thereby rendering them equally problematic for achieving success.

The exoteric is simply the theory and most outward aspect of the religious system. It covers for instance, the idea of Buddha and the four noble truths, or the 5 pillars of Islam. This stretches in theory all the way to the idea of how the inner soul of humans gains awakening by transformation through intense practice of the theory. The esoteric is the most inner aspect of the theory and path that includes understanding and experiencing the most subtle aspect of metaphysics, which would be the soul, heart, mind, and spirit. The term used in Western spiritual parlance for the spiritual path is metaphysics.

Metaphysics is a temporary law (science) designed by divine intelligence to return humans to their lost, or in Buddhist nomenclature, "Buddha" nature. The temporal metaphysical law is strictly existent because of the fact that humans have strayed inwardly from their original nature. If that had not happened there would be no metaphysics, religion, or thought of God!

The law is most important in that it is the only way to truth for a particular person as defined by their time and place in the cosmos. The law, though, is a science (law of cause and affect) of light hidden amongst the confusion of sciences of darkness. But what most religionists and philosophers do not seem to grasp is that we are talking about a science with a particular intent. That intent is to be a guide to return humans to their true nature. This science is supposed to be from "God" in Western mysticism, but for all our intent it could be defined as being from the truth. The truth is the method that works to get from point a, to point b, by a law of light within darkness. This "law"(metaphysics) has no intrinsic value, or reality outside its intent, that is why no reality has ever been given to it by God, or genuine mystics. Therefore we have different paths for different ages whose wisdom and intelligence is based on the knowledge of what is good for one person in their time and place. Any universality ascribed to this law by making it "evolutionary" is incorrect. The essential LAW OF METAPHYSICS IS A TEMPORARY LAW, THAT IS DESIGNED FOR A SPECIFIC PURPOSE, it has no intrinsic universality relative to evolution or anything other than the cosmic laws that deal with the return of something to its source of origin after a breach in its integrity. It is this breach - the ellipse - that is the driver behind the science itself and drives its consistency, not any thing evolutionary. Metaphysics is a science of methodologies relating to this return; based evolutionarily, strictly on the cultural time-centered personalities of the guiding archetype-intelligences that preside over this science by light: Abraham, Moses, Jesus, Muhammad, Buddha, Chrisna, etc. These names and others unmentioned are the essence of the science relative to the time period of their culture. They are the sciences (knowledges) in fact embodied in the personalities and epochs of these individuals. The proposition by mystics, like Muhammad, Buddha, and Moses, that none of these sciences are eternal is another indication of the temporality of true mysticism.

The more accurate and descriptive term - Evolutionary Intelligence - I relate to the concept of intelligence that knows the spiritual needs within time and place of all epochs, and is about all anything evolutionary applicable to metaphysics that we are going to get. This, though, is truly an aspect of "God" or the divine (in spiritual technical terms in Sufism this science is called "Nur (light) al Muhammad." In Christian mysticology it is known as the logos.)

WHAT WE ARE DOING IN SPIRITUAL PRACTICE

We are fundamentally trying to reach from the standpoint of the two lower worlds, the worlds of the mind and body, the worlds of the two higher realms - the world of the spirit and soul. That is what all spiritual practice is trying to do, whether we know this or not.

These two higher worlds cannot be reached directly by us because we are locked in the lower consciousness of the mind and the body because of the ellipse.

All metaphysical practice is a psycho-spiritual attempt to reach these realms of the soul, spirit, and affect them, as well as trying to raise consciousness to perceive them.

Prayer, meditation, etc., is like taking medicine for a headache through the mouth to affect the brain. This is not a direct analogy because the mouth and body are in the same realm as the brain. That tells us how difficult it is to reach the higher realms of the spirit and soul from our lower dimensional standpoint of the mind and body. We are using practices from the mind-body standpoint to affect the spirit-soul realm.

We are trying to affect this realm because the source of our separation - the ellipse - is in that realm. These spiritual practices are supposed to be medicine to heal this sickness that exists in the higher realms of the spirit and soul, where the ellipse presides.

Only with a genuine metaphysic theory, that is practiced ardently and sincerely, by the science of light, of the reality of the law, not emotionalism, dogmatism, or ethnocentric sectarianism, does one become successful in his or her spiritual endeavor in terms of the metaphysical yardstick of enlightenment, awakening, and then completion.

This is very important to understand. Just reaching these two higher realms is not the point in metaphysics alone. The point in metaphysics is to heal the aspect of these higher realms that causes us to be in this imbalanced condition of suffering and fragmentation of our consciousness.

Traveling to the higher realms without the medicine to heal oneself does not do us much good.

Being a spiritualist, a seer, a psychic, or one of those other people who have reached and commune with the two higher worlds of the spirit and soul using certain Shamanistic practices ... all this does not mean they have remotely perfected themselves, even though reaching these realms is a good indication of progress on the spiritual path. Remember, the metaphysical philosophy is about healing, returning to our holistic Buddha nature, then on to nirvana in Buddhism and Yoga, or, in Sufi parlance, completed human; Christ consciousness in Gnosticism.

Essentially, metaphysical practices are about humans reaching these higher realms through spiritual exercises in order to affect the healing of the realm, so the entire being can return to the balanced nature of perfection, and completion of our holistic self. It is the energy coming from the lower realm through spiritual practices that reaches the higher realms, effecting healing and hastening the return of the lost circles to the essence.

When Gautama Siddhartha (Buddha) said *"I am awake,"* he had through deep constant spiritual practices - particularly samandi, and insight meditation - along with practicing the virtues, reached the two higher realms. As well, he healed them of the ellipse as the source of samsara (delusion) and the veils to reality. However, remember this, he healed the ellipse by the perfect practice or methodology of his spiritual ways - by the reality of the law of cause and affect. Moreover, his exercises worked as a medicine, and a tool for spiritually moving the energy produced from his practices to the two higher realms of being, in the spirit and soul, and effectively erasing the corrupt ellipse structure. Proper spiritual practice works just like a medicine does on a sickness, there is absolutely no sky-God, Santa Claus, Wizard of Oz type input from any external "God" that helps this process.

This samsara is the descriptive word that describes the condition of our delusional perception that is a result of the ellipse in the soul-essence that causes it.

Consequently, practicing the contemplative sciences with all one's heart and ability, as well as practicing the virtues, by the grace of the law, one may find great success on the spiritual path.

If one is able and capable, one should certainly follow the principles and guidelines of one of the metaphysical paths to perfection. If an individual is not capable of following the path perfectly, then the advice is to follow the virtues as best one can, keep away from major, and as much as possible, minor corruptions and deeds. This followed by doing as much one can in the meditative sciences.

It must be understood clearly that not all are capable of following the path all the way to completion. Therefore, a believer in the spiritual laws, who doesn't as yet have the ability to be successful on this path, should by the law of the virtues become the best human being they can until God, nature or whatever opens up a way for them to tread the spiritual path with a possibility of success.

These paths include the major and some minor mystical ways such as Sufism, Islamic Sufism, Yoga, Vedanta-Advaita, Zen Buddhism, Buddhism, Taoism, Kabbalah, Gnostic and Mystic Christianity. That is, generally, any path that has at least an understanding of the theory of enlightenment. Also, a combination of the best that remains in these paths is also an option since most of our great mystical teachers from Jesus to Muhammad, Moses, Lao Tzu, and Buddha never were adherents of the religions of their day. They each reformed the paths of their day with a renewed vigor, intelligence and guidance that eschewed the literalist sectarian dogma and error that the existing religions had fallen into. Any religions that shun sectarianism, racism, dogmatism, literalism, sexism, or what has become known as fundamentalism is a good path for return to perfection. It is advisable to shun the dark shade of the mind referred to above and anything that degrades or belittles, for any reason, any group of people beyond the most horrendous evildoers. Additionally, shun like the plague paths that exalt themselves over others, or claim a special nearness to God. In essence, one should take a middle course and emphasize the best and most humane loving aspect of the faiths and paths of our day. This of course is difficult because the paths have been polluted by clerical dogma, philosophical additions, ignorance, and over-emotional sentiment for a long time. Therefore, one must be a wise, courageous, and mature person to etch out from these paths and religions a positive harmonious and loving way of returning to the perfect circle - our true nature - even with all the difficulties that will face us. Not to overuse the cliché, but love is the key to return to God, along with tolerance, wisdom, courage, compassion, patience, and most importantly, the love of knowledge and disdain for ignorance. Those are some of the important provisions on the path.

ESOTERIC LAW OF ALCHEMY

Three steps and two fathers

We will now discuss real physics in this section - that is the esoteric subtle nuances of this theoretical spiritual system. I will use one of the most fundamental laws in metaphysics - the path of return to our lost nature out of the ellipse - to explain these inner phenomena, the esoteric law Jesus expounded in his statement, "I and the Father are one." This means, on the highest level of interpretation, that the aspirant has to merge into the law he is trying to understand and practice. There are three major steps, or stages in the mystical path, and a merging with the father. This merging with the "father" actually has two aspects to it: one is the aspect of general metaphysics, and the other is the aspect of the meditative sciences. I will call the first aspect of metaphysics the first father, and the second aspect, of the meditative sciences, the second father. I will elucidate on the three steps, two fathers concept in explaining this metaphor mystically.

Merging indicates genuine change to a higher nature. The father here is the science, knowledge that works to make the inner changes. The Sufi saint and martyr, Hallaj, illustrates this in the famous aphorism that got him murdered by religious bigots, centuries ago. He said, "I am truth." He had merged with the metaphysical law - thereby changing his nature, and his inner sight. Also, he had only begun to merge with the higher father that would have taken him all the way to God.

Any human can become "truth" by as best they can reflecting it, through the heavenly realities of - first - the temporal law of metaphysics - father one - which is all about merging in order to create movement or change in the heart; then becoming that essential "Buddha" nature, or perfect soul (enlightenment). Then further to the real goal of the spiritual path - completion.

Becoming truth on one level is merely seeing the reality that we are already truth. It is only the veils of the heart that are sourced in the corrupted essence - *ellipse* - that veils our inner eye from truth. Think of that: you carry truth with you every second of the day, we can only reflect the truth; it is an impossibility not to. It is only that the reflection you see is polluted by veils of the heart that makes us see delusion as truth. Everything is truth, but our wayward perception prevents us from seeing this completely, cognitively.

These are the general three steps-stages that are included in the classical mystical way included in all genuine paths. Traveling by the law by merging with the law-alchemy/transformation, then arriving at the reacquisition of the true nature - Buddha Nature - of self, and then ascending to the final stage that the Sufis call completion - Isanul-Kamil.

An important technical distinction in metaphysics is that literally one does merge with the temporal metaphysical law (first father) that is generally understood in terms of ordinary exoteric and esoteric religion: following the virtues and meditative arts indicates a literal transformation of the heart. This is a true merging, as Jesus' statement indicates; but the other half of this is not a literal phenomenon, and relates to the meditative sciences that are the aspect of merging with the higher father (second father) or the transcendent outward Adam. What is this outward (divine) higher Adamic (father)? If you remember in two earlier chapters we went over the 6(7) natures and explained them all. The higher or transcendent Adamic nature (the nature that reality is in a science) is the outward father that we employ in metaphysics as a major tool to reach enlightenment.

In Sufi cosmology, the story of Muhammad who rode a winged horse named Buraq in his ascension to heaven to talk to God, is a metaphorical allusion to the higher external Adamic nature we utilize in returning to God. Buraq, the horse Muhammad rode to heaven according to Islamic tradition, I interpret as a symbol for the meditative sciences. Again, this tradition is allegorical, not literal. However, it is literal in the sense that Muhammad inwardly ascended to the highest heavens on his meditative power, not really a physical horse.

The Buddhist concept of the Bodhisattva (one who delays complete enlightenment, until all sentient beings are saved) however it has been thought of, is reflective of the second step, merging with the law - first father - that being the bodhisattva "delaying his enlightenment," the one before completion that technically can be called enlightenment. But is not the final step. It is the condition of perfection that existed before the breach of an aspect of the soul-essence that befell humans from their true nature; as mentioned above.

Similarly, in Sufi cosmology the concept of Fana (annihilation of the personality) of all human traits precedes the final and complete experience of annihilation of anything other than God in the person, which is Baqa (merging with God, by shaking off the religious personality). Fana can be described as the annihilation of the conditioned or lower self into the higher self (father one) and Baqa can be described as the effacing of that higher self into the divine (father two). Both of these two different figurative references to the second and third step in awakening in Buddhism and Sufism are not just poetic references, but two literal descriptions of two major stages of awakening.

These are very subtle points to distinguish:

Step one is the merging of the personality with the metaphysical law that allows the person to merge with the first father. This is the opening up of the personality to change.

Step two is the aspirant shaking off the personality of the first father (metaphysical temporal law) and merging with the second father (external Higher Adam (meditative sciences).

Step or stage three, the final stage, is the seeker in a sense eschewing the second father - the practice of meditative sciences - in order to arrive at ultimate truth as nothing, before God.

By "God", what do we really mean? Of course, in Buddhism, this concept is deemed irrelevant, but there is a correspondence in Sufism, Vedanta, and some forms of Buddhism in the notion that God is truth, or reality. So when we say all personalities and selves must be effaced before God, substitute truth, or reality for God and you will understand this concept better.

This above reference to effacing "personality" in truth is the source of the Buddhist doctrine of no self. The error of some Buddhists in interpreting this doctrine is them replacing soul with self - personality. There is without a shadow of a doubt no reality behind any conditioned personality, but on the other hand, the soul is not conditioned or created.

Lest this concept is confused, again: truth, God, or reality ultimately has no personality, in any form whatsoever. A personality or "self" is only a temporary, created, or conditioned self that is pretty well destined for eternal effacement before God or truth. It is the feeling self, knowing self, or being self, outside of any personality self, that is real.

That is the true meaning of the concept of Buddha's no-self doctrine, as well as the Sufi concept of annihilation or Fana/Baqa.

As for the idea of becoming one with the science (metaphysical law), that presents a challenge itself for the seeker at the second stage to the third stage or from Fana to Baqa or from the bodhisattva to full enlightenment in Buddhism. For, as suggested above, even the perfect nature must be annihilated in God so all traces of self no longer remain, and one is consumed by nothing but the truth. For the perfect nature or Buddha nature, (father in heaven) is a science or knowledge (phenomenon) just like anything else that has to be effaced at the threshold of ultimate reality, so true being or completion can be.

Additionally, in God, or reality as we have shown, there are seven transcendent natures so anything else of a nature is transitory-creation. By transcendent I mean something that exists, and has always existed, and always will. The seventh nature is the holistic self.

The existential reality of this description of spiritual movement exists on the psychic and physical level of being, the first two worlds in metaphysical cosmology.

The return of the nine elements that are missing from the essence as I described above in Chapter 4 (the cause of our fall) operates on the third and fourth level of the soul and spirit level.

Four worlds
4. SOUL
3. SPIRIT
2. MIND
1. BODY

The Ellipse: The Fall and Rise of the Human Soul

The soul is the highest world as relates to us, it is the highest "holon" - the holistic self. This is the container of the three elements spoken about in Chapter 11: The Tao, Holy Spirit, and the Essence. So whatever spiritual activity we do on our gross level - prayers, meditation, virtues, etc. - we are attempting to alter the configuration (doing metaphysics) of these most subtle elements of our soul, an aspect of which (the essence) has been corrupted. On the other hand, when we perform the mundane, we at least reflect this soul existentially, but we are not then trying to consciously affect that deepest realm or change it. That is what metaphysics is in theory supposed to do. The existential problem of this physics is that this subtle realm is difficult to affect or reach from the standpoint of the two worlds of the body and mind we are so familiar with. All this work done from the lower mind and body level is to affect or accelerate the natural healing of the ellipse corruption that has affected the essence.

Mr. Alan Kavel, the very astute writer, esotericist, and essayist informs us in one of his internet essays that metaphysics literally means: after the physical, a practical reference to the organization of Aristotle's books. That may be so, but surely the science of synchronicity informs us that such a mundane reference to this has a higher meaning, and that meaning is to penetrate by psychic sciences, the other dimensions of the two higher worlds, this essentially is all what metaphysics is trying to do.

A general summation of the path of the mystic, which exists solely to heal the ellipse, begins with the method of the metaphysical idea usually performed through one of the religious paths: Buddhism, Sufism, Vedanta, Gnosticism, and others. This idea is the attempt at inner alchemy through merging with the "father" or law; this law is a created science (that can transform the heart) specifically to deal with the existential corruption of the essence. This merging must be genuine in the sense that the individual has to accept the criteria of the law's reality: ethics; morality; correct mentality; methodologies, purity of heart. This must be a true merging with the aspirant's heart of the fact of this law that will offer the student legitimate change; by change is meant something far beyond what one might think, for this change is not merely a different set of attitudes about life, or anything that superficial.

At this stage, certain dangers are apparent as history tells us many mystics have been overwhelmed by this condition and have faced serious problems. This is so because the very integrity of this condition has to merge with the understanding of its temporality. This is not the truth stage. Remember the path is temporal; at this side of the truth, the aspirant must shed the cloak of temporal science (metaphysical law) that has taken him only to the threshold of truth, that he can only enter, finally, with a different set of "laws" than he has understood.

The Sufi tale of the seeker at the door of truth can only begin to convey this reality: It seems that after 40 years of seeking, Mahmud finally entered the truth realm, and he knocked at the door. Behind the door it was asked, "Who is it?"

"It is I," said the seeker.

"Go away," said the voice behind the door. A year later, our aspirant returned and again knocked. Again, he was asked, "Who is it?"

"It is Mahmud," said the seeker.

"Go away," said the voice behind the door. Years later, after further searching, the aspirant returned, and knocked at the door of reality, again as before, a voice behind the door asked, "Who is it?"

The seeker said, "It is thou." The door was then opened to him.

Mahmud at first did not realize that in order to enter the door to ultimate truth, he had to shed his new personality - created by merging with the law of return, which became his new I, that he thought would be enough to enter. He learned there is no I but God, nor is there room for any personality-self, however lofty it is, in the truth.

In fact, approaching the door to truth, as the tradition of the Sufis concerning Muhammad's Miraj or ascension to heaven illustrates the same principle as Buddha's famous aphorism: "When over the river, discard the boat." The tradition is that Buraq, the mystical horse, did not enter heaven with Muhammad. Therefore, that which was discarded is Buraq or the transcendent nature that the personality rode on to the truth, which in this case is the meditative sciences. This also tells us conclusively that man cannot really existentially merge with God, but can only merge with his own, inner "God" to know reality. Remember, man merges with the law, not God. Merging with the law only means that genuine change has taken place in the heart of the seeker.

Also, this scenario confirms the idea that ultimately there is no self-personality. All self-personalities are phenomena, or conditioning.

As for Buraq, not being allowed to enter heaven with Muhammad is the essence of the Buddhist aphorism, "When over the river, discard the boat."

The boat here is Buraq, the symbol of what Muhammad rode to enlightenment, which is to say the meditative sciences, which at awakening one needs to discard since the inner psychic apparatus of the one enlightened no longer needs the meditative sciences.

It is as one Sufi savant said: the Sufi at this point becomes mysteriously in union with the intent of God.

SUMMARY

Two laws

The temporal metaphysical law is the personality the aspirant MUST merge with, (he becoming the law, in a sense), and is the first aspect of the father. Merging means essentially adapting inwardly - morality, ethics, and virtues - indicating a permanent transformation of the inner self.

Parenthetically speaking, for those who think the cultivating of virtues is subjective, or not necessary, then what they do not understand is that this cultivation is guiding the developmental aspect of the path in a positive direction. This development has to be guided one way or the other or the development can be negative.

This metaphysical system is specifically created to deal with the fall of man; it has no valid application outside of this syndrome.

The Higher law - the law of one of the transcendent natures - called the law of the higher Adam is the second aspect of the father that he rides on to "God". This is God itself in a sense, because it is a transcendent reality. Thus, being one of the six natures has an existence in macrocosm as well as microcosm. Though in this case, is an outer vehicle being used temporarily by the fallen human. The transcendent law becomes for humans the practice of the meditative sciences. This is done with the hope of traveling on this nature - to God (reality/truth). This outer higher Adam cannot be merged with literally; we, in a sense ride on top of this law to God, as Muhammad rode on Buraq to heaven. What occurs is that (practicing the transcendent Adamic law, by doing meditative sciences) re-tunes humans with their (own) inner higher Adam. This is a very fine point to emphasis here, and should be thought about carefully in order to understand these principles.

Then, as Buddha says, "When over the river, discard the boat", the boat being the meditative sciences. For we no longer need the external higher Adamic law when enlightenment is attained. It is not literally a part of us physically (remember it is the macrocosmic higher Adam we were riding on - Buraq, or practicing meditative sciences) because we have our own inner higher Adam. This inner higher Adam at perfection does not have to practice ritualistic transformative meditation because the seeker has reconnected with the smooth running inner higher Adam of his own being, which does not require training.

At this point we begin to see reality as it is, not as it is not, as we see now through the veils of samsara (delusion).

To see is the fundamental goal of the mystic path. To see reality as it truly is, is the goal of Buddhism, Mystic Christianity,

Kabbalah, Sufism, Yoga, Taoism, and all genuine paths to the truth and fulfillment.

To see is the basis of true enlightenment, because one cannot really know, be, or do without it.

To see is dependent on traveling the entire path that realigns the inward spirit with the essence, itself, so all veils of darkness are lifted from the inner eye of the soul, and reality is seen for what it is, and accepted by the heart of the individual. That acceptance is dependent on this realignment because it brings the heart to submission to the dharma, thereby bringing genuine inner peace to the soul. This submission is based on the perception of reality. The heart wavers in anxiety, fear, and grief because it reflects the condition of the soul before it reaches peace. This is the greatest veil to truth, and seeing reality as it is extinguishes this veil. That is why the "religion' of God is based on peace and submission. Not submission to an angry vindictive sky-God, but submission to universal law, or in Buddhist terms, dharma. By submission, I mean the transcendent concept of acceptance of how the universe works.

THE CIRCLES RETURN

Using one of the genuine mystical paths, one may return to the perfect nature and free oneself from the ellipse by one's Herculean deeds. However, it is very challenging and almost impossible for an established ellipse. I hesitate to offer this pessimistic view but will not for the sake of spiritual correctness be disingenuous here and tell people what they want to hear. History and the condition of the human race is testimony to this statement. What is in our favor is the certainty of the cosmic clock, in that as the ellipse structures become less necessary by the edict of the time clock of the EI, more elliptical beings will be able to resolve their selves and return to perfection. The present day population explosion seems to indicate an intention of the EI to complete the affair of the ellipse, in order that we can move on to our destiny of reunion with our perfect and complete nature. The general path out of the imperfect circle is nothing new as expressed above. The only unique element I offer here to contribute to general understanding is a precise description of the traumatic events, inward, and outward, of the fall of man, and of a renewed awareness of the imminence of divine perfection and beauty that is open to all.

One should genuinely accept a religious theory of a balanced way of the exoteric (outer) and esoteric (inner) path that teaches love, unity and the doing of good to one's fellow beings and the entire universe, with an understanding and belief in the theory of enlightenment. Eventually, if done with real intent and right action (spiritual meditative exercises, and good deeds) the two missing circles that were ejected from our soul at the fall begin to return at an accelerated pace, for by nature they are very slowly returning to the sphere they were ejected from. (Without a successful metaphysical practice, this return is extremely slow). This only begins the path of return. If the circles are to remain on a smooth trajectory of return to the essence, then right practice must be maintained, or they will lose their path of return, eject themselves back out of orbit, and produce a very rocky experience for the seeker. Therefore, it is necessary to maintain a sincere practice of healthy spiritual action that reestablishes the trauma-ridden inner soul to its past position of a perfect circle. For the two missing circles will get reestablished when a consistent practice of the moral and spiritual paths is maintained. Then what occurs is the repairing of the circuitry of the circle - rivers of light - that was damaged at our fall in primordial times. This takes time, and the eventual goal is the returning of the perfect orbit of the soul's essence. Then the perfect energy-processing machine (the soul) reignites. Once the circuitry is repaired, the angelic circles are simultaneously seen by the inner eye of the soul again. At this point the consciousness energy of the being, after reestablishing the operation of the 4 circle-corresponding to the Fana stage in Sufism, and the bodhisattva in Buddhism, returns to the source 5-circle and reestablishes perfection - return to the source God

(Figure 21). That is the same condition as the transcendent Adam at this point, the altitude we fell from in antiquity *(Figure 22)*. This has to occur for full enlightenment to happen, that is returning to the 5-source garden. (Remember we had no business in the 4-circle from the standpoint of the 5-circle in the first place, for that created the ellipse, the source of all of our problems.) This is a level of enlightenment, though not full enlightenment, because there is another final enlightenment stage to go, which I will describe in a minute. Therefore, mystically returning to the 5-source circle, reestablishing our original perfection is obligatory.

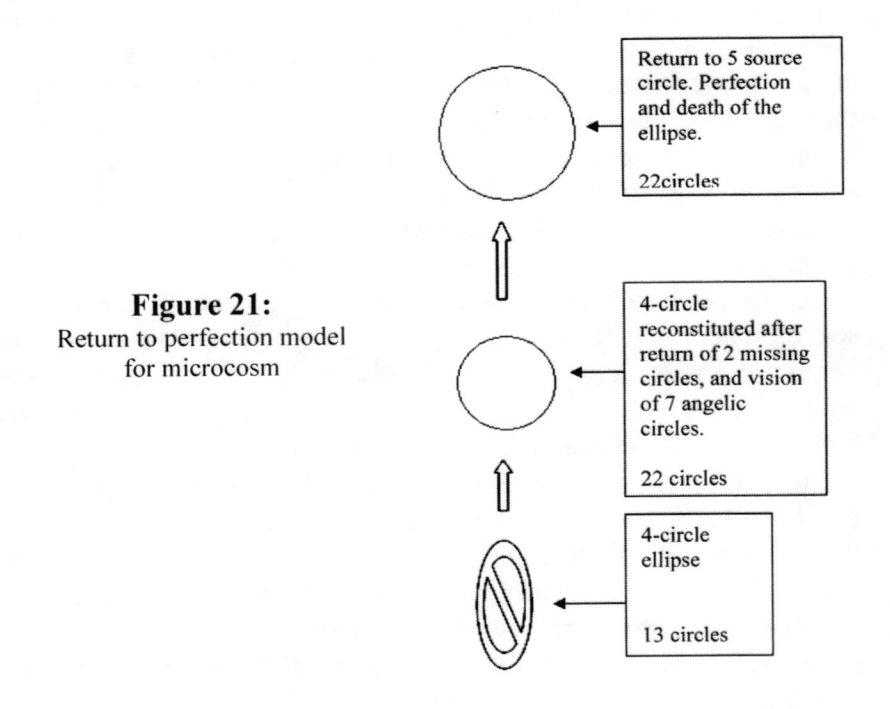

Figure 21:
Return to perfection model for microcosm

- Return to 5 source circle. Perfection and death of the ellipse.
  22circles
- 4-circle reconstituted after return of 2 missing circles, and vision of 7 angelic circles.
  22 circles
- 4-circle ellipse
  13 circles

That is a degree of enlightenment, and essentially the death of the ellipse, as well as the return to the perfect circle, the original Garden of Eden. From that point is a further growth - full enlightenment (completion) of the soul (Atman), which has to do with the perfect alignment of the essence (circles) and the Holy Spirit, which produce expansion or resurrection *(Figure 23)*. From there our destiny in the beatitude of paradise states has no limit. This is the mystical and divine alchemenical transformation of the divine soul (circles) to a steady completed divine heart of infinite beatitude of unlimited pure energy. The famous light verse in the Quran (24:35) describes so wondrously this state and its divine elevated station: as it describes the final configuration of the completed soul, Atman, the God-consciousness-being:

"Allah is the light of the heavens and the earth. A likeness of his light is a pillar on which is a lamp - the lamp is in a glass, the glass is as it were a brightly shining star-lit from a blessed olive tree, neither Eastern nor Western, the olive whereof gives light, though fire touch it not, light upon light. Allah guides to his light whom he pleases. And Allah sets forth parables for men, and Allah is knower of all things."

Allah is the light of the heavens and the earth, refers to the reality that God is the light of the microcosm and macrocosm, or the light of the above and below. The light represents the Rau-unbound infinite spirit. The pillar referred to is a symbol for the Holy Spirit, which rests at the seat of the essence (circles). The lamp is the circle essence structure itself, the glass is the final expansion of the 5-circle, which at resurrection becomes a physical contour that outlines the circle structure, then 21 circles surrounded by (glass) outer 5-circle expanded, or resurrected (brightly shinning star). A blessed "olive" refers to the Tao which gives "light" (brings in energy from the Rau) the blessings of circular travel. Neither Eastern nor Western is the perfect balanced condition of the soul at this point. The olive giving light is the pure radiation of the circle energy that comes from the tree. Fire touches it not, is the reality that only pure uncorrupted uncreated energy is manifested by the circle radiation.

Light upon light

Essentially, the verse - interpreted at this high level - is a description of the God-state of the macrocosmic and microcosmic being-entity, or what is termed as the God-being. This is the highest station-state of humans who are now divine, abiding in the completed soul (paradise states) in perfect alignment with the Holy Spirit after resurrection. This is the Nirvana of Buddhism, Atman of Yoga, Insanul-Kamil (completed person) of the Sufis, and Christ-Consciousness of Gnosticism. Here begins our existence in the state of pure energy consciousness. Moreover, the individual has a unique experience and understanding of the immense power of the 4-circle line of descent, and how it operates. The potential for enormous creative ability, and the possibilities for immensely positive and beautiful experiences of pure science, creativity, and joy that is nothing that we have ever experienced or could imagine. That does not even cover the joy and consciousness beatitude that exists in the pure uncreated energy states in the other lineages, in the trees of paradise consciousness.

So there we have it, a simple roadmap to return to perfection. This theoretical roadmap of course does not begin to describe the energy it takes to do what is described above, or the difficulties one will have in carrying out this work. Indeed, make no mistake about it, success at this point in cosmic time is extremely difficult, and this path is challenging, for us as ellipse beings. Do not diminish the reality of this difficulty in treading this path. The power of the ellipse is enormous, and is indicative of the awesome might of this soul machine, that we are only a shadow of its splendid condition in perfection and completion. This difficulty is measured here against success, not measured in terms of the middle course path (slow as you go) that is a viable alternative since any spiritual progress can contribute to a person's success.

Chapter 15

The End of Time: Perfection and Completion

The Prophet Muhammad is reported to have said: That God said to him when asked,
O Lord what can contain you? The lord said: Nothing in the heavens and the earth can contain me, but the heart of my believing servant.

The spiritual technical term "perfection" refers to the condition of our souls before the creation of the ellipse, the fallen man. There is a memory trace of this primordial condition on our psyche, which is why we can discuss this today, for as the saying goes "There is nothing new under the sun." This is why on one level of the mystical Sufi path they emphasis the practice of memory exercises that mystically attempt to jog the memory back to the primordial, perfect Adamic archetype in our souls. In every human being, there is the archetypical trace of our primordial perfect selves - exemplified by the Quranic emphasizes on the constant theme of "return to us." This is only a metaphor for the return to the perfect inner higher Adamic stage, and the next more important step, from the 4-circle to return to the perfect 5-source circle - God *(Figure 22)*.

Then on to the resurrection-expansion stage which, as mentioned above, is the "marriage or perfect alignment of the essence (circles) with the Holy Spirit," which compels expansion of the 5-source circle. This is the highest stage a human can reach *(Figure 23)*.

This Holy Spirit which the Quran says: "God breathed in to us at creation" is that universal cosmic spirit that the Hindu and Tantric mystics refer to in their cosmological doctrines. This Holy Spirit, in uniting with the essence of God (circles), and the

Perfection:

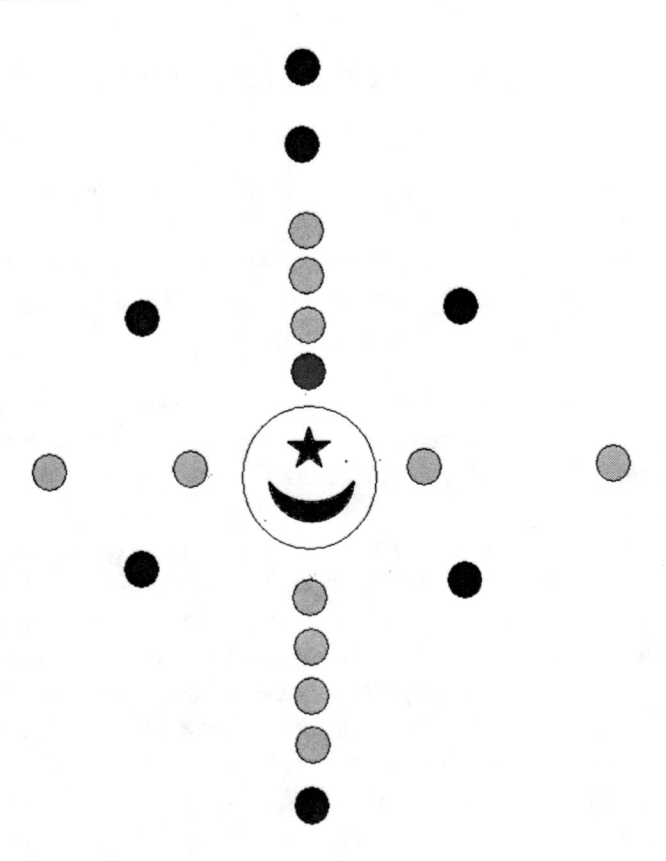

Figure 22: Essence in perfection

marriage of these divine unit energies in the human being, is the ultimate return or union with God, referred to in all of our mystical and religious philosophies. This is the end of our "creation" or completion. The unit being is then alive, for our inner, energy-producing soul-spirit is in a state so sublime as could be technically classified as beyond perfection.

The resurrection is literally an expansion of the 5-source-circle to encompass the entire garden-circle of paradise states, where the inner circle will still then be the chakra circle, but the lataif circle will fill the place of the expanded five circle.

Completion:

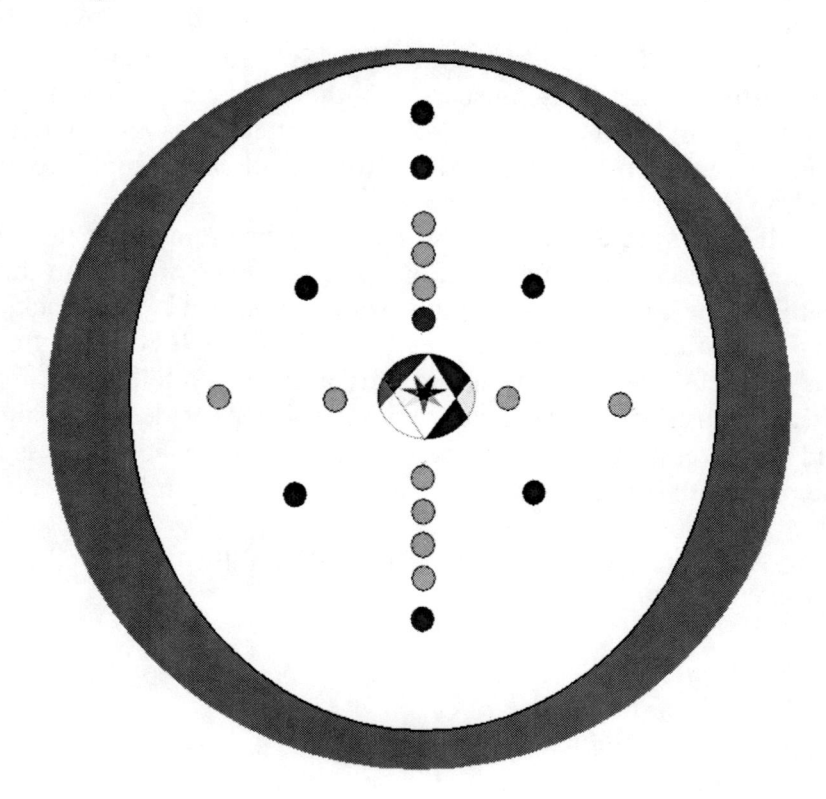

Figure 23: Depiction of completion of essence. The colors on the seven-pointed star are the chakra colors, and the colors outside represent the colors of the lataifs. The darker circles are the archangel essences (upper circles) that number seven. The lighter circles are the lower circles, as the darker circle above the new axis circle, is the 4-circle, the forbidden tree. Behind this *Holy Essence* - the face of God - is the Holy Spirit, the Tao, and beyond that is only the infinite Rau.

The resurrection then has two aspects to it: the return of humans to perfection, by returning to the 5-circle and retiring forever the ellipse. The second stage is the expansion of the 5-circle to preside over the essence - the completion of the Garden of Eden. This is the completion of the *Station of the Soul:* the evolutionary enlightenment of this New Age era, Insanul-Kamil of the Sufis, Christ-consciousness of the Gnostics, Nirvana of Buddhism and Yoga. Nothing else will alter the soul from this point, though the Rau energy will always nourish the soul through the Tao in eternity.

All beings are evolving to the station of the divine soul; there is nothing that can prevent this. For this is the will of the non-dual reality of the cosmos. It will guide us to completion by the clock of the end of time. Therein, the great veil of tears will be lifted from our souls, by the hand of the great Evolutionary Intelligence, that inter-dimensional divine guide, whose spirit pervades all things, and whose essence is the source of all states, and whose love is at the root of all being...

AT THAT TIME WE ALL WILL TRULY FEAST... AT THE BANQUET OF DIVINE UNITY

For then
God will show us all
Something beautiful
"For whence he wills a thing...he only says to it:
"Be

and it is."

Annotations

Secret Knowledge

On the next page is a Google earth picture of the ellipse circle [imperfect circle-negative energy] in the flesh, right behind the White House! The White House is shown above the ellipse circle in the picture.

The ellipse is the negative energy reality of our microcosmic soul that came into existence [EVOLVED] after the fall of man. It exists as our solar system - THE ABOVE. It also exists in all of our inner selves - THE BELOW.

This picture of Ellipse Park in Washington DC is a synchronistic reminder to us of our inner and outer non-harmonious existence. Noting that since it is related to the government of the greatest temporal power on earth, the US government, this indicates something that is not hard to deduce. This level of the ellipse - in an occult sense rules us - is shown in the gross macrocosm that we LIVE INWARDLY AND OUTWARDLY INVOLUNTARILY IN. *The ellipse is the real Matrix.*

The reason we have metaphysics, religion, good, evil, and suffering is the ellipse. We are attempting to regain the inner PERFECT CIRCLE [we possessed in the Garden of Eden] THE ANTI-THESIS of the Ellipse, by practicing religion, and mysticism.

The Ellipse: The Fall and Rise of the Human Soul

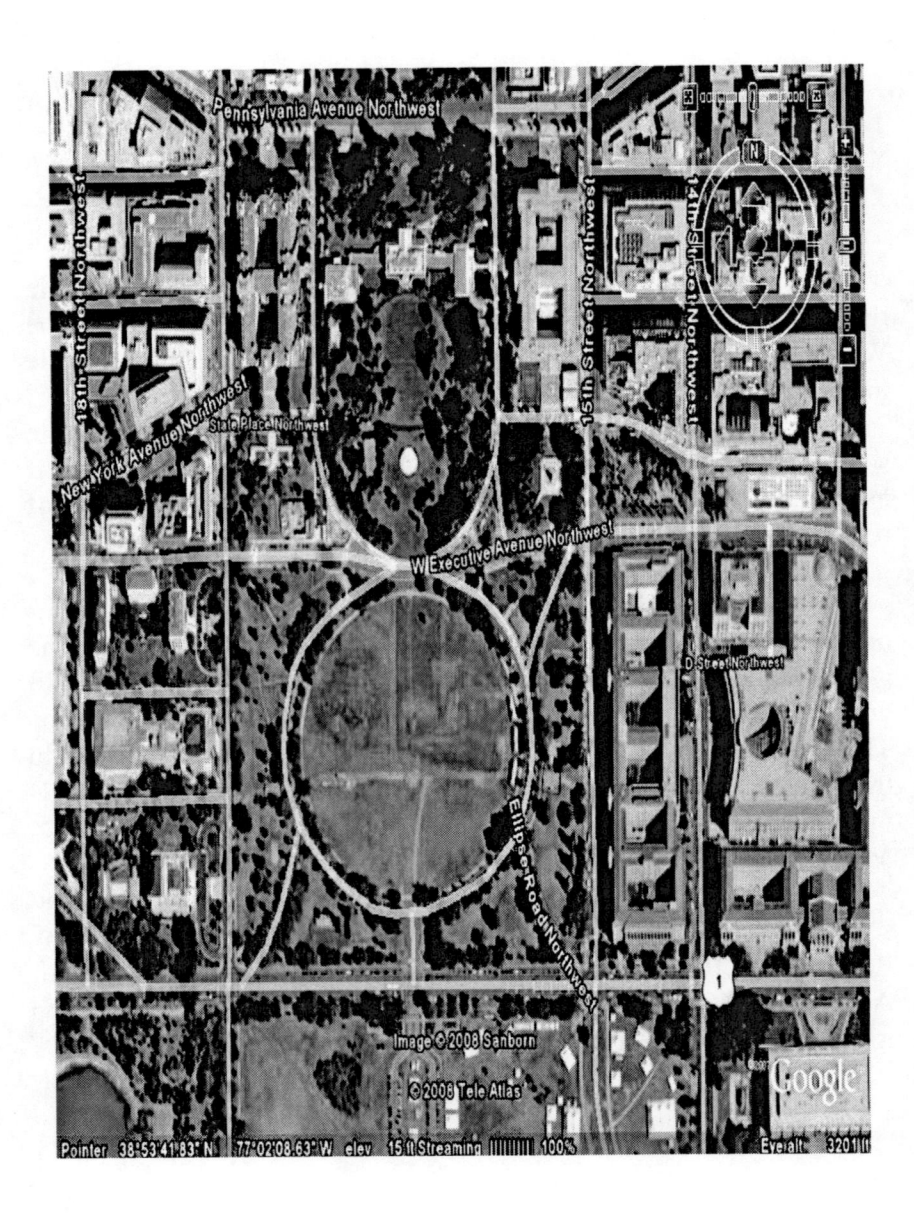

Annotations

Number Places on Perfect Circle

[Essence]

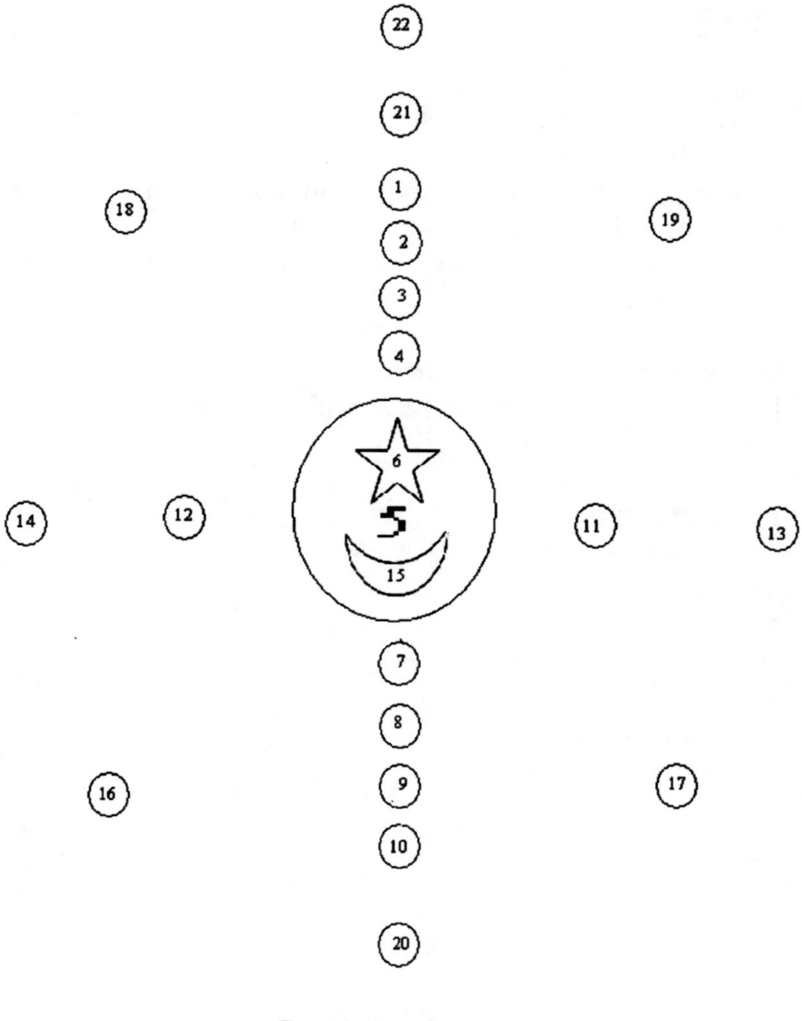

�22

㉑

①
②
③
④

⑱   ⑲

⑭   ⑫   ★6 5 ⑪   ⑬
          ☾15

⑦
⑧
⑯   ⑨   ⑰
⑩

⑳

Garden of Eden

Secret of the Cosmos

When we look out in the vast cosmos we are seeing inside ourselves - expanding outward infinitely as a direct reflection of our inner.

Every solar system a station
Every sun a balance
Every planet a state
Billions and billions of years from now, when this station dissolves will not end the eye of the essence
For she will merely descend in another station to abide for another eternity

By then some of us will realize
That we are but dust in the station of the mistress
Animated by the state of her
To move along her spirit
To the end of time
When dust becomes
Another sun
And it says
I have never not been here
And that sun becomes
Another mistress
Of the spirit

Annotations

Unity Of Faiths

BUDDHA (The Awake)

Buddha is wise because of his understanding of human nature; that whatever lofty metaphysical truths we grasp, one must grasp the truth of his or her own state or condition.

The essence of it is: how we feel, what we are and what we do.
How we feel: we are free of suffering
What we are: we are awake
What we do: we do good, that frees us from suffering because we are awake.

**Muhammad
on whom be peace**

The seal of the Prophets

Muhammad is the seal of the prophets because he in his life reflected all the essences of God equally at the same time. Thereby serving exquisitely the Supreme Arch-angel.

His nine wives were the Holy Spirit structure or the Kab-Allah, as well as the Tau - His Is-LAM.

ISIS-LAMA the supreme **Goddesses** along with the LAMA, supreme teacher.

So we have all the major religions and mystical paths unified by this combination: for example, Isis is also the Virgin Mary as also indicates the ancient Egyptian religion the source of Judaism, Christianity, and Islam.

The Lama word brings in Hinduism and Buddhism, culminating in the essential religion of humans in the word Islam that denotes submission to the Goddess (God/ESSENCE) and peace.

JESUS
son of man
sun of God
son of Mary

He was the son of man as myth representing the fall of all humans from grace.

The crucifixion is all of humankind's sojourn in the darkness of separation from God (ellipse).

His resurrection represents the inevitable resurrection of all humans.

He was sun of God representing the center of the solar system, the 12 disciples circumnavigating around him.

And son of the Virgin Mary after being baptized by John the Baptist.

As well as the son of The Mary Magdalene (may God perfect her) as the SUN of God. or the sun of this solar system.

The Virgin Mary "Mother of God" was the reflection of the light of the central sun, the source of our solar system.

In cosmic metaphysics, to be the son of these divine Goddess energies means only that they Jesus, and John,

Were captured by their celestial lights.

Thereby becoming the moon to their sun.

As they, the Goddesses, are a moon to God.

THE GODS

All religions are basically the same, only differing in nomenclature.

The great Hindu religion accused of being "pagan" because of the reference to many Gods is the most rich and versatile creed on the planet. And far from separating, the Gods have enriched the concept unlike any creed. Most astute Hindus acknowledge the unity of God or his essential one-ness but also acknowledge God's rich complexity of the many faces of the divine that interact with the human experience. So indeed it seems that the Gods are equal to or the same in fact as the Muslim 99 attributes of God, or the famous trinity of Christianity in the Brahman, Vishnu, and Shiva of our great Hindu tradition.

THE FAR EAST

Be it the formidable people of the land of the rising sun, or the noble people of China, our Far Eastern heritage is unrivalled by anything in history. As in the case of the Nipponese, for instance, exhibit such natural perfection don't seem to need a religion!

These originators and inheritors of great wisdom traditions, such as Shinto in Japan, Tao and Confucianism in China, and both peoples contribution to Zen Buddhism, seem to illustrate that their enduring abilities are an example that the applied wisdom has worked in the case of the people of the Far Eastern lands.

The wisdom traditions that sprung from the great culture of China - Taoism and Confucianism - are two of the world's premier legacies of wisdom that aid in guiding humans to ultimate perfection.

Lao Tzu, the subtle master who always reminded man that he should be more of a woman if he wants ultimate success in life, is one of our greatest sages, who left the consummate knowledge that sprang from the Tao tradition, a tradition pregnant with knowledge and sciences.

THE PEOPLE OF GOD

The myth of God's people has always existed in one form or another in the heart of man. This is so because the reality in the macrocosm [in a sense] is that GOD has to speak to one person - or entity - at a time in this elliptical world by order of its reality. So that in every era, indeed, there has to be a "chosen race, people, or person." This archetypal reality is reflected in our great Jewish tradition that has brought the world some of the greatest, most righteous, wisest, men and women in the world. It is not their Jewishness alone that makes them formidable people; it is their association with this principle that helps to sustain this.

Unfortunately our Jewish people have played a price down the ages for this, which only goes to show us that with greatness goes suffering and pain. Out of this cauldron of pain, though, people from Moses to Einstein have brought the human race an unparalleled heritage of greatness, knowledge and wisdom that has served as a model of unique sustenance throughout the ages for us all.

THE RELIGION OF LOVE

"I profess the religion of love,
Wherever its caravan turns
Along the way, that is the belief,
The faith I keep"

The great Sufi poet Rumi ruminated on his religion. Although a cleric and traditional Muslim, the renowned mystic always elevated in his teaching the universality of religion and mysticism. Certainly the Christian, Jew, Muslim or Pagan was welcomed by Rumi as a student. This is indicative of the culmination of the evolution of our religious sense. Indeed the mystics of all the great faiths are without a doubt the inheritors of the mantle of grace that is the key to the salvation of man.

Mediaeval Kabalists, Sufis and Gnostic Christians - probably just like today - sat around a fire and shared their mystic secrets with each other, notwithstanding their different faiths. No one taught humanity this but the mystics. There would be no interfaith dialogue, integral movement, world-wide unity conferences, without the spirit and culture of this handed down to us from the great adepts of our holy inner traditions: Taoism, Buddhism, Zen Buddhism, Sufism, Mystic and Gnostic Christianity, Kabbalah, and many people who have traversed across the spirit of all these schools of thought.

THE NATIVES

Thanks to the much-maligned period called the New Age, our Native creeds from the warm jungles of Africa to the dry steppe hills of South America or the ancient caves in the heart of Europe have experienced a renaissance that has resurrected forgotten traditions that have enriched us all. The Shaman medicine man like the Druid priest holds a place of esteem in our culture as never before.

The native priests and priestesses of all of our cultures were the first pioneers who dared venture in the murky waters of the soul and were the foundation of ALL the faiths. Indeed, whether it be the call to prayer in our mosques, the chants of the Christian and Buddhists monks, or the hymns of the Kabalists; the source of it all was some lonely place in a jungle, a cave, or a mountain where some brave soul began

THE JOURNEY BACK TO GOD FOR US ALL

AMEN

THE END
OF TIME,
IS THE BEGINNING